AF839290

World Soils Book Series

Series editor

Prof. Alfred E. Hartemink
Department of Soil Science, FD Hole Soils Laboratory
University of Wisconsin–Madison
Madison
USA

For further volumes:
http://www.springer.com/series/8915

Aims and Scope

The *World Soils Book Series* brings together soil information and soil knowledge of a particular country in a concise and reader-friendly way. The books include sections on soil research history, geomorphology, major soil types, soil maps, soil properties, soil classification, soil fertility, land use and vegetation, soil management, and soils and humans.

International Union of Soil Sciences

Olafur Arnalds

The Soils of Iceland

Springer

Olafur Arnalds
Environmental Sciences
Agricultural University of Iceland
Borgarnes
Iceland

ISSN 2211-1255 ISSN 2211-1263 (electronic)
World Soils Book Series
ISBN 978-94-017-9620-0 ISBN 978-94-017-9621-7 (eBook)
DOI 10.1007/978-94-017-9621-7

Library of Congress Control Number: 2014955324

Springer Dordrecht Heidelberg New York London
© Springer Science+Business Media Dordrecht 2015
This work is subject to copyright. All rights are reserved by the Publisher, whether the whole or part of the material is concerned, specifically the rights of translation, reprinting, reuse of illustrations, recitation, broadcasting, reproduction on microfilms or in any other physical way, and transmission or information storage and retrieval, electronic adaptation, computer software, or by similar or dissimilar methodology now known or hereafter developed.
The use of general descriptive names, registered names, trademarks, service marks, etc. in this publication does not imply, even in the absence of a specific statement, that such names are exempt from the relevant protective laws and regulations and therefore free for general use.
The publisher, the authors and the editors are safe to assume that the advice and information in this book are believed to be true and accurate at the date of publication. Neither the publisher nor the authors or the editors give a warranty, express or implied, with respect to the material contained herein or for any errors or omissions that may have been made.

Printed on acid-free paper

Springer Science+Business Media B.V. Dordrecht is part of Springer Science+Business Media (www.springer.com)

Preface

Approach—Organization of the Book

Icelandic soils are quite different in nature from the soils found on either side of the Atlantic, because of the active volcanism, unique climate conditions, and especially active geomorphic processes shaping the island. This book on *The Soils of Iceland* takes a broad approach considering soils as a part of ecosystems. The book provides the fundamental background to Icelandic nature, including climate (Chap. 2) and the volcanic geology (Chap. 3), which is the single most important factor in determining soil formation on this island: the formation of Andosols and Vitrisols. The book also provides some basic information about the people and the use of the land (Chap. 2). Vegetation of the country is described in Chap. 4, based on a comprehensive land cover database housed at the Agricultural University of Iceland. The author also found it necessary to present a short review on Andosols (Chap. 5), which Icelandic soils basically are, with the often peculiar properties that characterize Andosols.

The more traditional soil science sectors are described in four Chaps. (6–9) on classification, physical properties, chemistry, and finally genesis and biological issues. Chapter 7 on physical properties suffers from lack of such studies in Iceland, but more is known about the basic chemistry and genesis. Information about land use is woven into these chapters, which include numerous photographs, tabular information, and figures representing various relationships between soil properties.

Iceland is a cold land with climate characterized by winter temperatures oscillating around zero. The soils are very frost susceptible leading to some of the most frost affected soils and nature anywhere outside of permafrost regions, which merits a special chapter (Chap. 10) on soil frost and its effect on Icelandic landscapes. Chapter 11 is devoted to the extensive aeolian environments, which influence all other ecosystems through redistribution of volcanic materials.

With land degradation as the major factor shaping Icelandic ecosystems, considerable effort is devoted to the history and processes of degradation in Chap. 12. And the question how is such knowledge achieved? Understanding soils is part of the literacy needed to read the land, to understand the state of the land, function, and condition; an approach taken in Chap. 12.

Punctuation—Icelandic

This book includes the names of many Icelanders, Icelandic place-names, and other words which involve the use of special characters when written in Icelandic such as á/Á, é/É, í/Í, ú/Ú, ý/Ý, and ö/Ö, which can be found in various other European languages. The characters also include ð/Đ, þ/Þ, and æ/Æ. All these characters can give problems in English publications, during printing and also with cross-referencing author names in modern-day reference databases. The author has selected to avoid the use of special Icelandic characters and hyphens in the names of authors or people except for the letter ö. Many other Icelandic authors have selected this approach, but the names can otherwise be found spelled in two or more different manners in databases. Thus, Ólafur Arnalds becomes Olafur Arnalds in this book. Þ is spelled Th in the reference lists, ð as d and æ as ae. However, Icelandic place names and journal names are spelled out in Icelandic.

Acknowledgments—The Roots of the Publication

I am greatly indebted to many people who have helped me to learn throughout my life and to those who took part in accumulating some of the information presented in this book. I am also thankful to those that have published research that is used in this book. In particular, I wish to name a few people, but I apologize for the length of this list, which does, however, give a clear indication of the roots of this publication.

Thanks to my mentors during graduate studies for their long-lasting friendship and encouragement: Jerry Nielsen (Montana State University), Tom Hallmark and Larry Wilding (Texas A&M University). Larry Wilding still continues to provide scientific challenge to his former student. Faculty, fellow students and friends in the US, in particular Steve Archer (formerly Texas A&M University now University of Arizona) and Paul McDaniel (University of Idaho). John Kimble (USDA-NRCS) helped me analyze and understand the nature of the Vitrisols, and Koji Wada (Kyushu University, Japan) helped with the mineralogy early on, together with my advisors (Hallmark and Wilding) and Joe Dixon at Texas A&M University.

Thorsteinn Tomasson, director of the Agricultural Research Institute and Sveinn Runolfsson, director of the Icelandic Soil Conservation Service (ISCS) provided encouragement and made it possible for me to initiate research on soils and soil erosion in Iceland after my studies abroad. I am also particularly indebted to my brother Andres Arnalds (ISCS) and other family members, who were instrumental in helping us through our studies in the USA.

The book draws considerable information from the Agricultural University of Iceland (AUI) soil database ("Ymir") that has been growing slowly but steadily over the years. Rannveig Guicharnaud (now EU Ispra), Bergrun Anna Oladottir (now University of Iceland), and Barbara Duran (Wageningen) were instrumental in soil sampling and analysis when a definite step was made to increase the database. Elin Asgeirsdottir (now Environmental Agency of Iceland), Sunna Askelsdottir (now ISCS) worked extensively in the lab. The lab has now, for many years, been firmly run by Brita Berglund (AUI). Baldur Vigfusson has been handling many of the instruments throughout this time, but Arngrimur Thorlacius (AUI) helped with establishing the soil analytical methodology. The database was the basis for the soil classification scheme presented in the book, in cooperation with Hlynur Oskarsson, Einar Gretarsson, and Fanney Gisladottir doing the GIS work to create the soil map of Iceland.

The European COST-622 Action (Volcanic Resources of Europe) proved to be very fertile cooperation. This book makes use of some of the wealth of information generated during this cooperation, and I am extremely grateful to my European friends including Francois Bartoli (INRA France, chairman), Peter Buurman Ed Meijer and AG Jongmans (Wageningen University), Paul Quantin (France), Eduardo Garcia-Rodeja (University of Santiago de Compostela, Spain), E. Fitzpatrick and Graeme Paton (University of Aberdeen), Otto Spaargaren (ISRIC, Wageningen), Georges Stoops (University of Gent), Folkert van Oort (INRA, France), Martin Gerard (IRD, France), Manuel Madeira (ISA, Lisboa, Portugal), Jorges Pinheiro (University of Azores, Portugal), Carmelo Dazzi (University of Palermo, Italy), Fabio Terribile (University of Napoli, Italy), Marisa Tejedor (University of La Laguna, Tenerife, Spain), A. Economou (Greece), Ward Chesworth (University of Guelph, Canada), and all the other participants in this COST Action. During the course of this work we met and exchanged ideas with various leading Andosol experts, including Randy Dahlgren (University of California, Davis), Paul McDaniel (University of Idaho), Roger Parfitt and Alan Hewitt (Landcare Research, New Zealand), and P.B. Warkentin (Idaho), all of which have provided inspiration through their scientific publications.

The author of this book has been a national representative to the European Soil Bureau, and has participated in European projects such as the *European Soil Atlas* and the *Soil Atlas of the Northern Circumpolar Region.* I am grateful to Luca Montanarella and his team (EU Ispra, Italy) for the cooperation and support throughout the years, and also the "polar-soils people", including Charles Tarnocai (Agriculture and Agri-food, Canada), John Kimble (USDA NRCS), the Russians, and all the others.

The participation in European projects dealing with soil protection in general provided some of the ideas presented in the book, including the SCAPE project and the cooperation with Anton Imeson (University of Amsterdam), Luca Montanarella (EU Ispra, Italy), Denis Peter (European Commission), and many others. Cooperation on various aspects of rangeland science is also evident in the book (e.g., Chaps. 4 and 12), and I am particularly grateful for the cooperation with Asa L. Aradottir (Agricultural University of Iceland), Ingvi Thorsteinsson (Agricultural Research Institute, Reykjavik), Johann Thorsson (Icelandic Soil Conservation Service), Steve Archer (University of Arizona), and Jeff Herrick (USDA/New Mexico State University), the Icelandic group at the Agricultural Research Institute and the Icelandic Institute of Natural History (Borgthor Magnusson, Sigurdur Magnusson, Gudmundur Gudjonsson, and many others), and Halldor Thorgeirsson (UN-FCCC).

Elin Fjola Thorarinsdottir (ISCS), Asgeir Jonsson (formerly ISCS), Fanney Gisladottir (ISCS now AUI), Einar Gretarsson (formerly AUI), and Sigmar Metuslaemsson (AUI, now Icelandic Institute of Natural History) were the backbone of the National Soil Erosion Assessment Project that earned us the *Nordic Nature and Environmental Award* in 1998. The Soil Erosion Database is extensively used in this book. The Nytjaland (Icelandic Farmland Database) was a continuation of the Soil Erosion Mapping Project, with the same people, but more recently with Björn Traustason (now Icelandic Forest Research) and Sigmundur Helgi Brink on board, and several other GIS people, but the Nytjaland database is also extensively used in this book, and is in part the background layer used for mapping soil properties and generating soil maps. I am grateful for the cooperation and friendship of all these people and others who have participated in the project.

The author thanks Jon Gudmundsson and Hlynur Oskarsson (AUI) for their friendship, data, cooperation, and fruitful discussions and for bringing the seriousness of the draining of Icelandic wetlands into international perspective. The AUI IGLUD database (Jon Gudmundsson) on land use changes is used considerably in the book. I am also thankful to Thorsteinn Gudmundsson (AUI) for discussions on the various aspects of soils and classification in particular.

Berglind Orradottir (AUI) was fundamental in gathering information about the physical nature of the soils in relation to restoration efforts (in part studying at Texas A&M University), and together with Johann Thorsson continues to provide insight and integrity to nature research. The graduate-level projects co-advised by the author of Berglind Orradottir, Rannveig Guicharnaud, Pall Kolka, and Birgir Oskarsson provided valuable soil information for the book.

The author thanks Fanney Gisladottir (AUI), Elin Fjola Thorarinsdottir (ISCS), Hjalti Sigurjonsson, Pavla Dagsson-Waldhauserova (University of Iceland/AUI), who all studied wind erosion during their graduate studies with the author (Chap. 11).

The author is thankful to the many people he has cooperated with in restoration studies, and in carbon sequestration research, mainly Asa L. Aradottir, Gudmundur Halldorsson, and Berglind Orradottir, but also Kristin Svavarsdottir, Pall Kolka, Gretar Gudbergsson†, Jon Gudmundsson, Johann Thorsson, Brita Berglund, and numerous others.

I am thankful to librarian Gudrun Thordardottir for her work and patience and to all the students who have survived my introduction course to soil science at the University of Iceland

and at the Agricultural University of Iceland. Many of the ideas, graphs, and figures of this book were developed for the course.

I thank Rector Agust Sigurdsson and Dean Hlynur Oskarsson at the Agricultural University, for giving me space and encouragement for writing this book. Thanks to all my colleagues at the Agricultural University and the Icelandic Soil Conservation Service for friendship and encouragement, not the least Gudjon Magnusson.

Many have been so kind to provide photographs and figures for the book, to all I am grateful but their names appear in the figure legends. Photographs not credited are from the author. Sigmundur Helgi Brink prepared numerous maps published in the book. Hlynur Oskarsson, Berglind Orradottir, and Bergrun Anna Oladottir helped me to improve some of the chapters of the book.

I am extremely grateful to Margret Jonsdottir, who has provided assistance to all of my major writing and editing projects and kept the books in balance for numerous years. She has helped me in handling many of the technical nuisances associated with running science projects, international conferences, and book editing.

Most importantly I thank Asa L. Aradottir, professor at the Agricultural University of Iceland and my companion in life and science for all the cooperation and for being there for us all.

Contents

About the Author

Olafur Arnalds has a background in Geology from the University of Iceland before earning an M.Sc. degree from Montana State University (1984) and a Ph.D. degree from Texas A&M University in Soil Science (1990). He has devoted his career from 1981 to Icelandic soils and geomorphic processes, but he has also been involved in soil conservation and land condition assessment, including ecological restoration and rangeland sciences. He first worked as a research scientist and subsequently as department head at the Agricultural Research Institute of Iceland, but later as a professor and a Dean at the Agricultural University of Iceland, where he currently is a professor.

Olafur Arnalds lead the National Survey of Soil Erosion in Iceland, which earned the *Nordic Nature and Environmental Award* in 1998, and the establishment of the 'Nytjaland'—The Icelandic Farmland Database. He has devoted research efforts in understanding the basic properties of Icelandic soils and geographical distribution of the soils, building a soil database for the country and a soil map for Iceland. His research efforts have also been centered on soil erosion, especially aeolian processes and dust production in Iceland.

Olafur Arnalds has written/co-written and edited/co-edited several international scientific books on soils, soil conservation, and rangeland sciences in addition to numerous peer-reviewed scientific papers (www.moldin.net). He has participated actively in Nordic and international groups on soils, soil conservation, and land condition. He has been active at several levels in nature protection in Iceland, which include writing educational materials on land condition in Iceland, participation in governmental committees and policy making, lecturing and teaching.

Abbreviations

Al_{ox}	Ammonium oxalate extractable Al
AUI	Agricultural University of Iceland
CEC	Cation Exchange Capacity
Fe_{ox}	Ammonium oxalate extractable Fe
ISCS	Icelandic Soil Conservation Service
Si_{ox}	Ammonium oxalate extractable Si
WRB	Word Reference Base, the IUSS/FAO Soil Classification System

Introduction

1

1.1 A Book on Icelandic Soils

Soils are fundamental for life on Earth, providing a media for cycling of energy, nutrients, and water. This vital resource is dynamic, ever changing with time as natural forces act on the surface, resulting in chemical weathering, modification of the biological activity, and formation of a great variety of soil types on the landscape. And then there is man, utilizing Earth for living; almost the entire Earth's surface has been modified by anthropogenic activities, often causing severe land degradation and even full ecosystem collapse.

The strong link between agronomy and soil science has sometimes placed restrictions of the realm of soil science (e.g., Arnalds 2006), while the "collaboration beyond traditionally defined soil science research disciplines" is increasingly emphasized by many (Adewopo et al. 2014; see also papers in Hartemink 2006). That is the very approach of this book: realizing soils as a part of the environment and ecosystems that are subjected to multiple processes that shape the landscapes of the Earth. Soils are subjected to degradation by various processes, a field of study that is a part of the soil science realm. Soil information is fundamental in understanding the condition of ecosystems; it is an important aspect of bringing soil information into a useful context (see Imeson 2012; also Adewopo et al. 2014), an approach taken here.

The nature that meets the eye on this island in the middle of the North Atlantic Ocean is unlike the nature we see on both sides of the Atlantic. The island has active volcanism, is subjected to more intense freeze–thaw processes than any other country, and geomorphic processes are both diverse and extremely active in places. Aeolian activity, with widespread redistribution of dust by wind erosion, shapes the character of all soils of Iceland. The soils are quite special, be it the fertile volcanic soils of the rich and pristine ecosystems, the unique blend of volcanic soils and Arctic bogs in the wetlands, or the unstable deserts that cover a large part of the country. And let us not be misled by the beauty of the landscape: most of the nature that meets the eye is very different from what our ancestors saw when they settled on the island; the impact of man has been dramatic.

The soils of the deserts merit special consideration, including the largest volcaniclastic sandy deserts in the world, dark in color due to the basaltic nature of the volcanic materials. The soils of deserts are given a special group name in Iceland: the Vitrisols, but these desert soils classify as Andisols under Soil Taxonomy.

This is a book on the soils of Iceland under the Springer series of the world soils. It is about what shapes the soils of Iceland and the soil landscapes, the services these soils provide, the severe impact of man, and how it is perceived. This book is intended to help the reader, the student of Icelandic nature, our numerous visitors—anyone interested in Icelandic nature—to understand the soils of Iceland and the processes that shape Icelandic landscapes.

The book makes frequent references to the different geographic regions of Iceland. These regions are presented here in Fig. 1.1 to aid the reader in locating areas named in the book. The figure also gives the names of the major glaciers, prominent features of the Icelandic landscapes.

1.2 Soil—Mold

In the Scandinavian languages, both Earth and soil are *jord*, in a similar way as the term sol is used in Latin languages. "Jord" is used to connote soil types just as "sol" is used as an ending for soil names under many classification systems such as the International Soil Science Union World Reference Base system (WRB) and the US Soil Taxonomy (as in Andosol under the WRB). However, in Icelandic, which in many ways represents the old Nordic tongue, the terms for soil are *jarðvegur* or the older more preferable term *mold*. "Jarðvegur", literarily meaning a soil-path or road, is apparently a translation of the Latin term "arvum", reflecting the plowed furrows, but "mold" appears very early in Icelandic texts and is gaining ground again in the Icelandic

O. Arnalds, *The Soils of Iceland*,
World Soils Book Series, DOI 10.1007/978-94-017-9621-7_1
© Springer Science+Business Media Dordrecht 2015

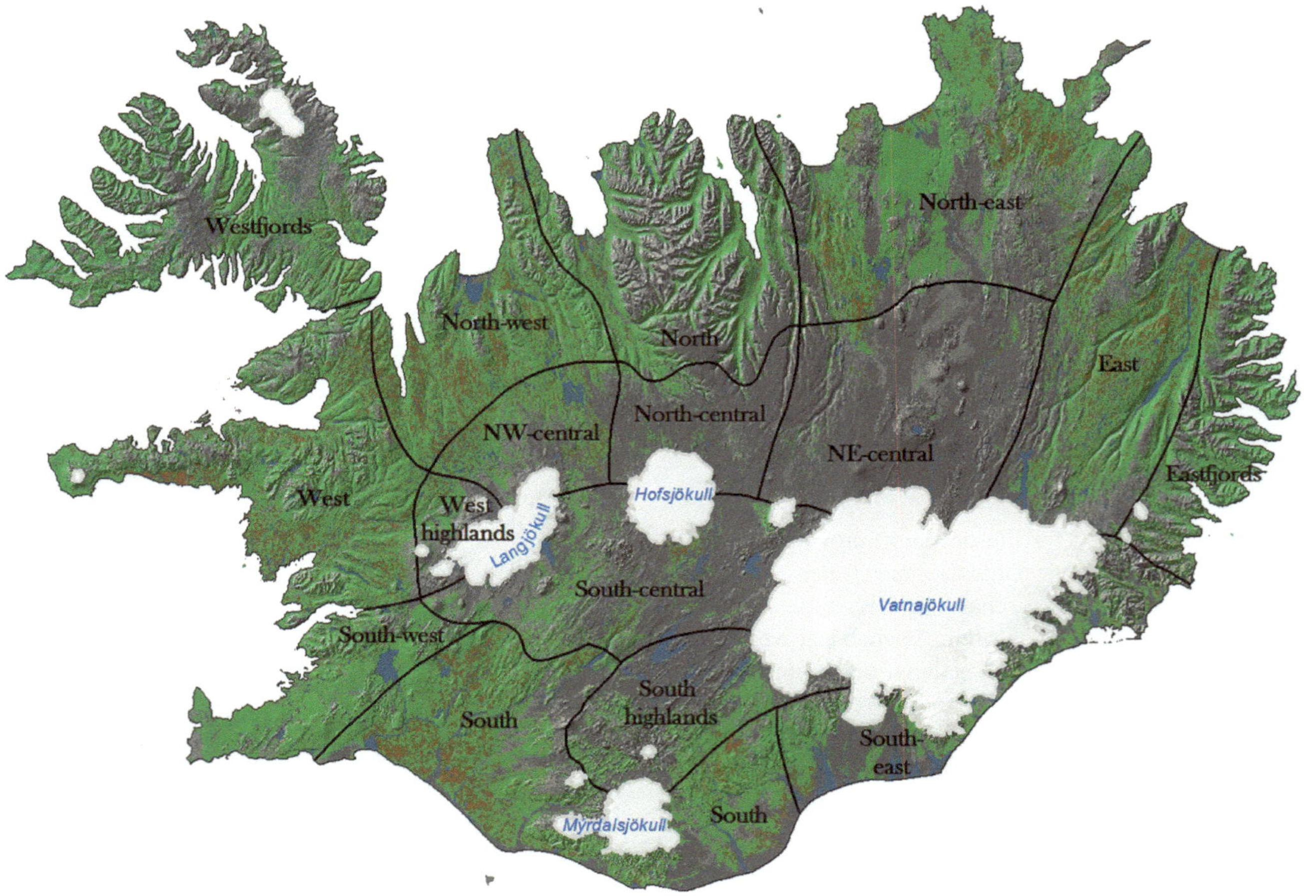

Fig. 1.1 The author's perception of the geographic areas in Iceland, which are referred to throughout the book

language. However, the term *jörð* is now used to coin the names of soil types such as "Brúnjörð" in Icelandic (Brown Andosols).

One has to bear in mind that agriculture in the sense of cultivating soils, which indeed was practiced during the Settlement time in Iceland, was not part of Icelandic farming throughout the Middle Ages, when the climate was not warm enough for cultivating common types of crops. That, in part explains the slow development of understanding of soils of Iceland.

1.3 Soil Science in Iceland

Soil science is in many ways a young discipline in Iceland, with only few scientists having received a Ph.D. degree in Soil Science, but several more have M.Sc. in Soils and related sciences. Interest is growing among students, often with emphasis on soil science as a part of the broader environmental sciences. The pioneers include Björn Johannesson (1960, 1988), who early on introduced a soil map and a book on the soils of Iceland. Gudmundsson (1994a) published a short textbook on soil science and made an attempt to adopt the present-day FAO classification for Icelandic soils (1994b). Bjarni Helgason published papers on soil formation in Southwest Iceland (1963, 1968) but his work included also research on the fertility of agricultural lands and soil nutrients (e.g., Helgason 2002). Fridrik Palmason pioneered soil nitrogen studies (e.g., Palmason et al. 1996) and there is a considerable volume of research devoted to grassland fertility and fertilization (e.g., Gudmundsson et al. 2004, 2005), which is in part cited under Chaps. 6–8.

The main work on soil science in Iceland has been undertaken by the Agricultural Research Institute of Iceland (Rala), which in 2005 became a part of the Agricultural University of Iceland (AUI), and this work is extensively used in this book. Many of the cited papers were published under the annual Agricultural Congresses in Iceland (Fræðaþing landbúnaðarins). Considerable information is also drawn for the purpose of this book from M.Sc. and Ph.D. works of students at both European and American universities, such as Thorsteinn Gudmundsson (Ph.D. University of Aberdeen), Rannveig Guicharnaud and Bergur Sigfusson (Ph.D. University of Aberdeen), Berglind Orradottir (M.Sc.

Texas A&M University), Birgir Oskarsson (M.Sc. University of Iceland), Pall Kolka (M.Sc. Ohio State University) and by the author (Ph.D. Texas A&M University). The author has been fortunate to participate in the education of young scientists in the field of soil science and environmental sciences, and their work is extensively used in the book.

The author has focused his research efforts on soil issues such as genesis/mineralogy, general properties, and classification with an establishment of a soil database (AUI Soil Database, heavily used in the book), but also surface processes that shape the Icelandic landscapes, land degradation, and land condition. Much information about the physical and chemical properties of soils in Iceland can be drawn from the joint European COST-622 Action (e.g., special issues of *Geoderma* (Bartoli et al. 2003) and *Catena* (Arnalds and Stahr 2004) and a book edited by Arnalds et al. (2007). Scientists at the University of Iceland, such as Gudrun Gisladottir and Sigurdur Gislason, have also contributed extensively to many aspects of Icelandic soil science, including the geochemistry and weathering. Research contributions in relation to the impact of man and degradation are numerous and include both Icelandic and foreign research efforts.

An attempt is made to cover all of this research by the author and others, in the subsequent chapters. The list above only gives an insight into the soil science in Iceland and origins of the information presented in this book, with more details given in the Preface in the section titled "Acknowledgements and the Roots for the Publication" and in the reference lists of each of the chapters.

References

Regarding punctuation and Icelandic characters in citations: See note on punctuation in the Preface

Adewopo JB, Van Zomeren C, Bhomia RK, Almaraz M, Bacon AR, Eggleston E, Judy JD, Lewis RW, Lusk M, Miller B, Moorberg C, Snyder EH, Tiedeman M (2014) Top-ranked priority research questions for soil science in the 21st century. Soil Sci Soc Am J. doi:10.2136/sssaj2013.07.0291

Arnalds O (2006) The future of soil science. In: Hartemink AE (ed) The future of soil science. International Union of Soil Sciences, Wageningen, pp 4–6

Arnalds O, Stahr K (2004) Volcanic soil resources: occurrence, development, and properties. Catena 56(Special Issue) (Elsevier, Amsterdam)

Arnalds O, Bartoli F, Buurman P, Oskarsson H, Stoops G, Garcia-Rodeja E (eds) (2007) Soils of volcanic regions in Europe. Springer, Heidelberg

Bartoli F, Buurman P, Delvaux B, Madeira M (2003) Volcanic soils: properties and processes as a function of soil genesis and land use. Geoderma 117(Special Issue) (Elsevier, Amsterdam)

Gudmundsson Th (1994a) Soil science. Búnaðarfélag Íslands, Reykjavik (in Icelandic)

Gudmundsson Th (1994b) The FAO classification system adapted to Icelandic conditions. Rala report no 167. Agricultural Research Institute, Reykjavik (in Icelandic)

Gudmundsson Th, Björnsson H, Thorvaldsson G (2004) Organic carbon accumulation and pH changes in an Andic Gleysol under a long-term fertilizer experiment in Iceland. Catena 56:213–224

Gudmundsson Th, Björnsson H, Thorvaldsson G (2005) Elemental composition, fractions and balance of nutrients in an Andic Gleysol under a long-term fertilizer experiment in Iceland. Icel Agric Sci 18:21–32

Hartemink AE (ed) (2006) The future of soil science. International Union of Soil Sciences, Wageningen

Helgason B (1963) Basaltic soils of South-west Iceland I. J Soil Sci 14:65–72

Helgason B (1968) Basaltic soils of South-west Iceland II. J Soil Sci 19:127–134

Helgason B (2002) Organic phosphorous in Icelandic soils. Icel Agric Sci 15:95–109

Imeson A (2012) Desertification, land degradation and sustainability. Paradigms, processes, principles and policies. Wiley-Blackwell, Chichester

Johannesson B (1960) The Soils of Iceland. With a generalized soil map. Atvinnudeild Háskólans, Ritröð B nr 13. Reykjavík. Available through the Agricultural University of Iceland

Johannesson B (1988) Íslenskur jarðvegur. (Icelandic soils). Reprinted version of a book from 1960 with extended appendix. Agricultural Research Institute (now Agricultural University of Iceland), Reykjavik

Palmason F, Thorgeirsson H, Sigurdardottir H, Björnsson H, Arnalds O (1996) Nitrogen mineralization in soil. Icel Agric Sci 10:185–208 (in Icelandic, English summary)

2 High in the North—Climate, People, and Agriculture

2.1 The Climate

Iceland, about 103,000 km^2 of volcanic island, is located far in the north, right under the Arctic Circle, between the 63° and 66.6° northerly latitudes, and 13–24° westerly longitudes (Fig. 2.1). In spite of the northerly global position, the climate is relatively mild. The reason is the oceanic climate and the powerful Gulf Stream, an ocean current that brings warm waters to the shores of Iceland. If it was not for this current, the climate of Iceland would be considerably colder than it is. However, a cold East Greenland Current influences the climate, but to the south the warm Irminger Current (part of the Gulf Stream) ensures relatively warm temperatures (Olafsson et al. 2007; Einarsson 1984). Low atmospheric pressure systems are frequent, sometimes referred to as the "Icelandic low" pressure system; this results in relatively high wind speeds, but Iceland is considered to be within the North Atlantic storm track (Olafsson et al. 2007). It is worth considering that continental areas at the same latitude as Iceland, in Siberia and Canada for example, experience much colder climates than Iceland, especially in winter. Mean monthly temperatures for Reykjavik on the one hand and several foreign cities are compared in Fig. 2.2.

The discussion in the subsequent chapters of this book often refers to geographic areas in Iceland, such as the North or Eastfjords. This division is presented in Fig. 2.3, but a larger version of the map was presented in the Introduction. It is based on the author's perception of this division, which is quite general, but may vary on how these lines are drawn between people and the purpose for each such map.

Typical average values for temperature and precipitation at several locations in Iceland are presented in Table 2.1. As can be seen from this table, winters in Iceland are relatively mild with temperatures commonly near zero (see also Fig. 2.2). Low pressure systems that bring relatively mild air masses are common in winter, and often raise the temperature above zero and bring ample moisture to the southern shores with southerly winds. There is more stable or longer lasting snow cover on the ground in the north and more so in the highlands. In the south, snow stays usually on the ground for a short time, for a day or a few days. Summers are, on the other hand, relatively cool compared to the neighboring countries (Fig. 2.2).

Table 2.1 presents averages based on the 1982–2012 (30 years) dataset, which results in significantly higher means than for the 1961–1990 averages, due to a warming trend. The averages for the last 5 years are still higher than the 1982–2012, an expression of global warming. The highest recorded temperature in Iceland is 30.5 °C (1939, Eastfjords), the lowest is −38 °C (1918, Northeast) (Olafsson et al. 2007).

There is noticeable difference in temperatures between the geographic areas of Iceland, but the elevation is the single most important factor affecting the mean temperatures (Fig. 2.4). As can be seen from this image, the mean annual temperatures are commonly 0–4 °C in the lowlands, but mostly 0 to −4 °C in the highlands. The highlands show Arctic character, where permafrost can be found in some locations under vegetation (see Chap. 10 on cryoturbation, also Thorhallsdottir 1997; Saemundsson et al. 2012).

A map of average temperatures in January and June is presented in Fig. 2.5. Winter temperatures are noticeably lower at the high elevations, especially in the north and northeast, but there are areas along the south coast where average temperatures are above freezing in January. Summer temperatures are also significantly higher in the southern and western lowlands (>10 °C), but are also favorable in the Eyjafjörður valleys (north) and inland valleys of the east. The warmest July areas are also those where barley production grows rapidly. It is interesting to note that the Arctic is sometimes defined as areas with average July temperature of <10 °C, but conventionally all of Iceland has been defined within the Arctic region (CAFF 2001, see Chap. 10).

Most parts of Iceland receive ample moisture for vegetation growth (Fig. 2.6) with >600 mm annual rainfall. Some areas in the south receive even more than 2,000 mm each year (record is >4,600 mm south of Vatnajökull glacier;

O. Arnalds, *The Soils of Iceland*,
World Soils Book Series, DOI 10.1007/978-94-017-9621-7_2
© Springer Science+Business Media Dordrecht 2015

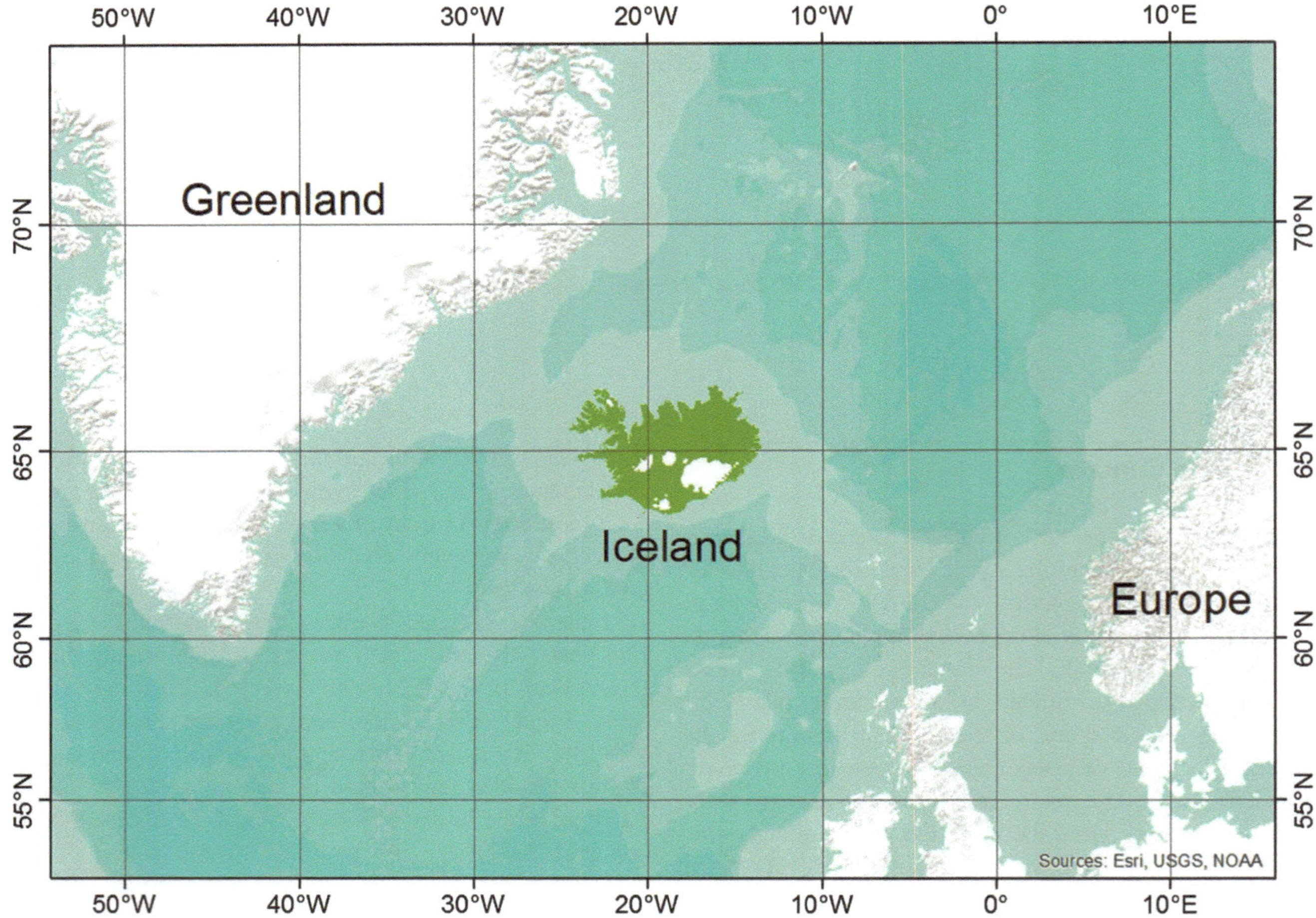

Fig. 2.1 Map showing the northerly location of Iceland. ESRI 2014; ESRi Basemaps

Olafsson et al. 2007). However, there are areas of low rainfall north of the Vatnajökull glacier with some areas receiving <400 mm rainfall. Humidity is usually as high as 75–90 %, but it can be quite low with cold dry air masses (see, e.g., Einarsson 1984). Rainfall is very common, but is often associated with long-lasting low intensity rainfall events. High intensity events do occur (>100 mm day^{-1}), especially in the south and in relation to passing of high-energy low-pressure systems (Olafsson et al. 2007).

Due to the common occurrence of low-pressure areas and periodic occurrence of storms (cyclons) blowing from the Arctic, Iceland is a windy country in general (Einarsson 1984; Olafsson et al. 2007). Wind speed can reach >30, and >50 m s^{-1} near mountains during severe storms, but wind speeds of 5–15 m s^{-1} are quite common (see Icelandic Meteorology Office web page, www.vedur.is on wind).

2.2 The People

Iceland was settled by Norsemen after 874 AD with rapid population increase during the first century. Iceland was an "independent" country (if one can say so about the states of the early Middle Ages) with a parliament, called "Alþingi" established in 930 AD, with neither royalty nor a king. Ties with Norway, however, remained close. The Icelandic Sagas were written mostly in 1150–1300 but their subjects often took place in the period 870–1000 AD (i.e., long after the events described). Early Icelandic scholars include Snorri Sturluson, who wrote both the history of the Norwegian kings (Heimskringla, covering 600 years of history) and an important mythological and poetic textbook (Snorra-Edda), which is a source for the majority of what is known today

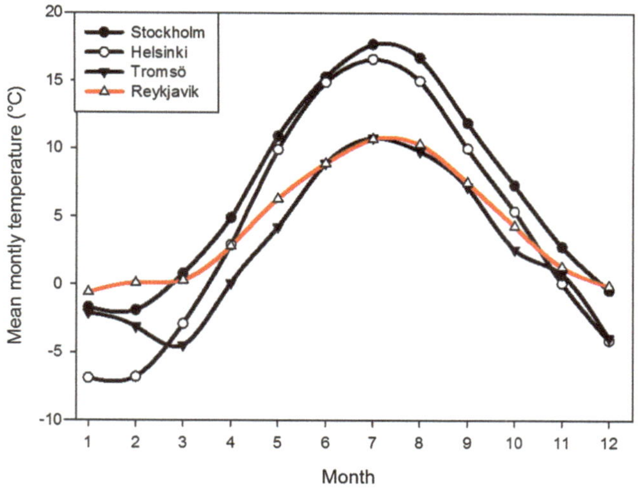

Fig. 2.2 Mean monthly averages for northerly cities. January and February are the mildest in Iceland (oceanic, Gulf Stream), but summer temperatures are considerably lower than in Stockholm (Sweden) and Helsinki (Finland), with more southerly and continental influences, but are similar as for Tromsö (69°\,41'\,N) at the northern coast of Norway. Data are from yr no, and may not represent same reference years. Using data from 1982 to 2012 (Table 2.1) gives about 0.5 °C higher means than presented here for Reykjavik

about the old German pagan religion. The early writings of the Icelanders, including the Sagas, provide an important glimpse into the natural history of Iceland at and soon after the Settlement. The Icelandic parliament is still called "Alþingi", and can be considered the oldest operating parliament in the world. Figure 2.7 shows the landscape of the old site of the parliament (Þingvellir, SW Iceland, UNESCO World Heritage Site).

The country was increasingly ruled by a few powerful families after the Settlement period, which led to a civil war during the thirteenth century and finally Norwegian rule (however, with loose ties) in 1262. Iceland came under Danish rule from 1662, which remained until the twentieth century, with partial independence achieved in 1904 and 1918 and full independence was declared in 1944, when Denmark was under German occupation. The population declined during the Middle Ages due to ecosystem collapse, civil strife, cooling climate, and natural disasters, which included the catastrophic Laki volcanic eruption in 1783 AD (e.g., Karlsson 2000). The population remained between 35,000 and 60,000 from the seventeenth century until the

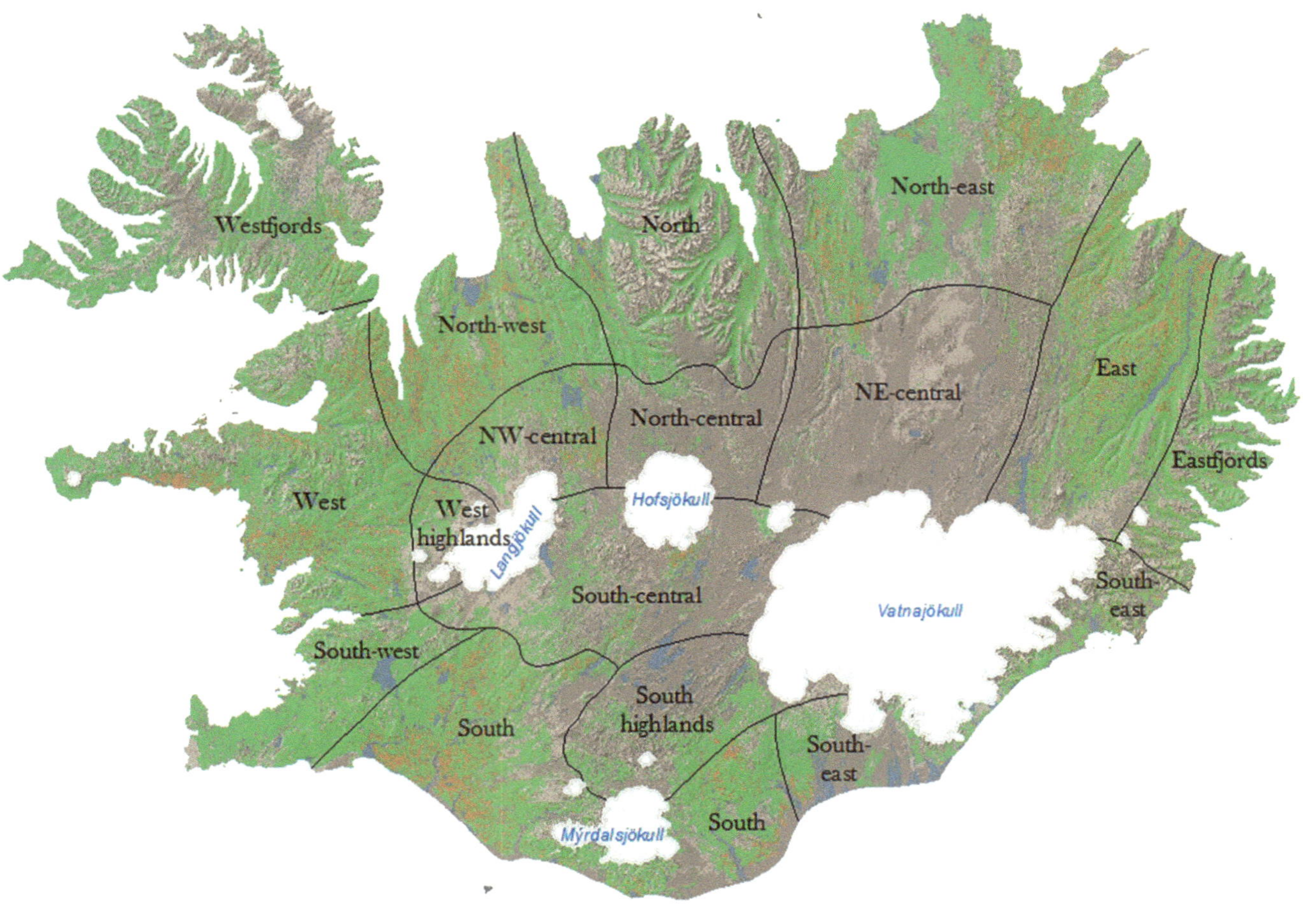

Fig. 2.3 The author's perception of the geographic areas in Iceland, which are referred to throughout the book

Table 2.1 Typical climate in Iceland during 1982–2002

Weather station (region in parenthesis)	Temperature (°C)			Precipitation (mm)		
	Mean annual	January	July	Mean annual	January	July
Reykjavík (SW)	4.8	0.2	11.3	868	87	53
Bolungarvík (W.fjords)	3.7	−0.1	10.3	812	85	41
Akureyri (N)	3.9	−1.1	11.2	559	63	33
Egilsstaðir (E)	3.3	−1.8	11.0	813	110	42
Kirkjubæjarklaustur (S)	4.9	0.2	11.6	1,775	168	126
Hella (S)	4.7	−1.3	12.5	1,258	119	92
Hveravellir (C)	−0.8	−6.0	7.6	754	74	52

Data from the Icelandic Meteorological Office of Iceland, Bolungarvík since 1994

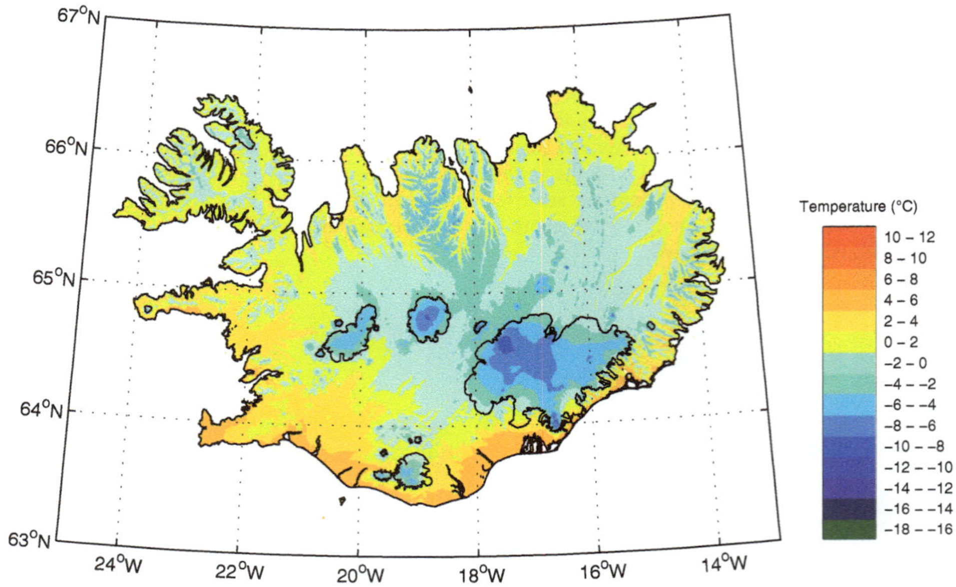

Fig. 2.4 Average annual temperature in Iceland during 1961–1990. Note that averages in Table 2.1 are for a different (warmer) time period of 1982–2012. Icelandic Meteorology Office web page (www.vedur.is), Björnsson et al. (2007a, b)

latter part of the nineteenth century, but rose rapidly during the twentieth century. Livestock numbers about 1700 AD were about 35,000 cattle, 280,000 sheep, and 26,000 horses (Karlsson 2000). The struggle for sustenance with limited winter fodder available for husbandry led to serious over-exploitation of land resources and soil erosion, as will be discussed in the last chapter of this book. The livestock numbers at about 1700 AD were not far from what they are today with about 75,000 cattle, 460,000 sheep, and 80,000 horses (see next section). Climate was believed to have been considerably colder during the Middle Ages than it is now, contributing to famine and ecosystem degradation (see the last chapter of this book).

Before World War I, most Icelanders lived in rural communities, on farms or small fishing villages. The country was largely controlled by landowners and many civil rights were limited to them: "... that it had been a dominant policy in Iceland over the centuries that all adults who did not run a farm of their own should serve as domestic servants in the homes of the farmers" (Karlsson 2000). Only landowners and government officials with very few exceptions were allowed to vote or to be voted to the parliament, with very slow change in the franchise until 1915 (Karlsson 2000).

Icelanders are about 320,000 today. About two-thirds of the population lives in the Reykjavik Capital area in Southwest Iceland. Agriculture is now only a small proportion of the total economic GDP (1.3 %) but a dominant factor in terms of control and use of land resources. Fisheries, industry (mostly smelters using hydropower and thermal energy), tourism, construction, and various services (including the financial sector) are the largest segments of Icelandic economy in terms of GDP (Statistics Iceland, www.statice.is). Tourism is growing rapidly.

Iceland has its own language, which has remained similar over the >1,100 years since the country was settled, which in part can be attributed to the isolation of the country during the Middle Ages.

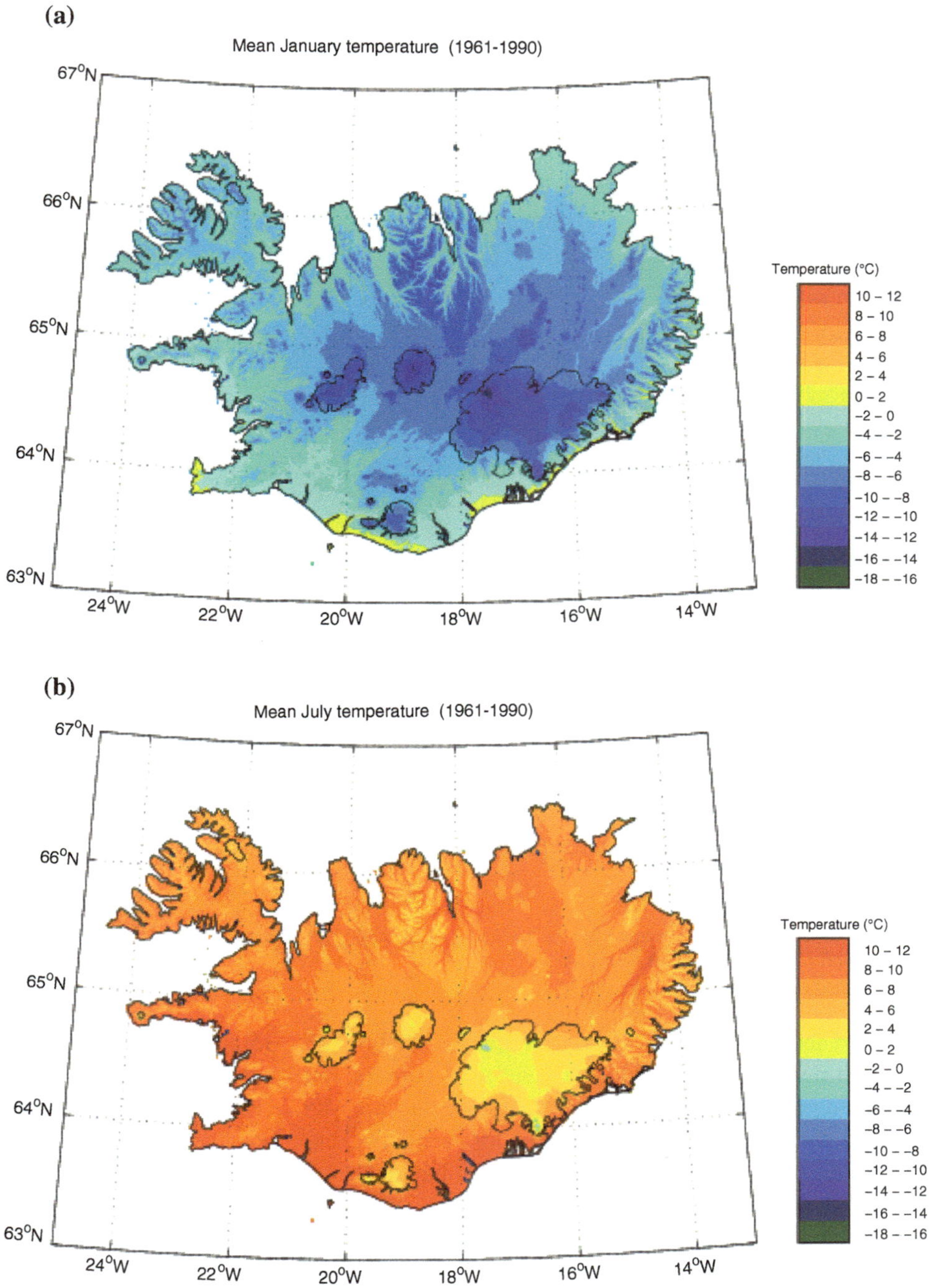

Fig. 2.5 Mean temperatures in January and July for 1961–1990. Note that averages in Table 2.1 are for a different (warmer) time period of 1982–2012. Icelandic Meteorology Office web page (www.vedur.is) and Björnsson et al. (2007a, b)

2.3 Agriculture and Land Use

Iceland is among the least inhabited countries of the world with about 3.3 persons per km^2 but the EU average is about 112 and Scandinavia has 15–22 inhabitant per km^2 (Helgadottir et al. 2013). Only one-fourth of the country is under 200 m elevation, where almost all of the population lives, bringing the population density in the lowlands to about 13 per km^2.

Active Icelandic farms are only about 2,600. The production, especially the dairy industry, can be considered grassland-based agriculture dependent on haymaking and grazing (see Helgadottir et al. 2013; Gudmundsson et al. 2013). Figure 2.8a, b exemplifies the Icelandic farming landscape. Dairy production and the production of sheep meat constitute the major proportion of the Icelandic agriculture (data from Statistics Iceland February 2013; www.statice.is), with about 26,000 dairy cows and 75,000 cattle in all, and 460,000 winterfed sheep. There are only about 1,000

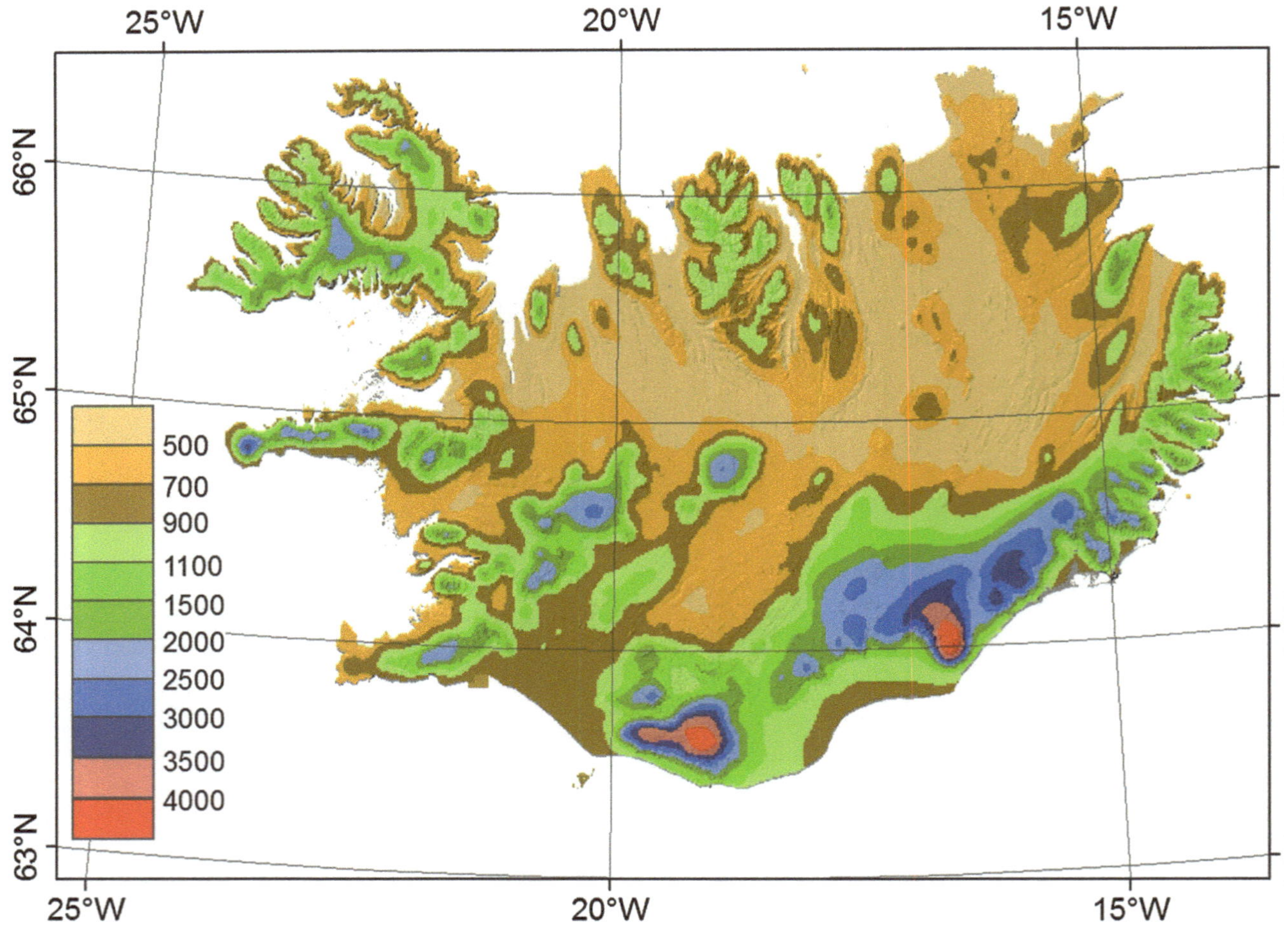

Fig. 2.6 Average annual precipitation in Iceland (1971–2000). Data from the Icelandic Met Office, Crochet et al. (2007), Crochet and Johannesson (2011). Prepared by SHB/AUI

goats and about 41,000 pigs. Horses are about 80,000, which is a high number compared to the number of humans, but horses are popular for recreation activities and they are also exported abroad. The average land area for each farm generally ranges between 500 and 2,000 ha, but hay fields are usually <100 ha, with 30–60 dairy cows and/or a few hundred sheep on each farm (Agricultural University Database).

The Icelandic livestock is in many ways unique, as these breeds, which were brought to Iceland 1,000–1,100 years ago, have remained in isolation in Iceland without mixing with other breeds during this time, maintaining an old important genetic pool (Adalsteinsson 1981; see also Helgadottir et al. 2013). The livestock are unusually colorful, and the Icelandic horse (pony) possesses some unique gaits that make it a valuable international commodity and a popular riding horse (Helgadottir et al. 2013). More than 150,000 horses of the Icelandic breed are found in other countries than Iceland (Helgadottir et al. 2013).

Animals, except horses, are kept indoors during winter, which calls for haymaking and storing fodder for the winter. Cultivated land is 1,100–1,300 km^2 (Nytjaland data; Helgadottir et al. 2013), but this number includes mostly permanent hay fields, which are plowed every 5–10 years. Cultivars commonly used in the hay fields include timothy (*Phleum pratense*), meadow foxtail (*Alopecurus pratensis*), smooth meadow grass (*Poa pratensis*), fescues (*Festuca rubra, F. pratensis*), and perennial ryegrass (*Lolium perenne*) (Helgadottir et al. 2013). Indigenous species are also common, including smooth meadow grass, red fescue (*Festuca rubra*), tufted hairgrass (*Deschampsia caespitosa*), and bentgrass (*Agrostis capillaris*), especially in old hay fields (Helgadottir et al. 2013). The soils of the hay fields are

Fig. 2.7 The medieval parliament site at Þingvellir, a UNESCO World Heritage Site. The area is characterized by active tectonic faults, with the depressions used for campsites and keeping horses in medieval times. *Photo* G.Kr. Johannesson

quite variable, but a large proportion is occupied by drained wetland soils, mainly Gleyic Andosols and Histic Andosols, but also Brown Andosols (soil classes explained later in the book).

Haymaking (Fig. 2.9) involves fertilizer amendments commonly in the range of 80–120 kg N ha^{-1}. Total national use of N is about 10,000 tons (www.statice.is). Due to the andic nature of the soils in Iceland (see chapter on Andosols), there is a need for P fertilizers in Icelandic crop production, (about 3,100 tons in total annually; www.statice.is). In addition to hay, barley is grown on about 5,000 ha, and other crops include annual ryegrass and oats. Barley production is growing rapidly with improved cultivars and a warming climate (see Helgadottir et al. 2013).

Both the sheep and dairy production are heavily subsidized by the government (for sheep: about equal to the income for selling the meat), and the subsidies rate among the highest in the world. Furthermore, most sheep farmers amend their income by employment off the farm.

The dairy cows are grazed on good grazing land close to the barn during summer. The sheep are set free on open rangelands, which include much of the highlands during summer, generally from late June or early July to early September, when the sheep are gathered. They are left unattended, as there are no predators that feed on the sheep except for occasional killings by the wild fox (Arctic fox) when the lambs are young.

The highland communal areas with the least vegetation cover and with severe erosion problems have been judged unsuitable for grazing after an extensive survey by the Agricultural Research Institute and the Icelandic Soil Conservation Service (Arnalds et al. 1997, 2001). Grazing of sheep in the highlands has caused considerable controversy, but attempts to close some of the poorest highland grazing

Fig. 2.8 **a** Farming landscape in central North Iceland, the wide dots are bales of hay. **b** A beautiful location for a farm in South Iceland. *Photos* Askell Thorisson, AUI

lands have failed (see Arnalds and Barkarson 2003; Crofts 2011). Icelandic nature conservation NGOs and the OECD (2001) have been very critical of land use policies in Iceland, and especially the poor state of the highlands. The highland grazing of the poorly vegetated communal areas constitutes only a minor part of the total sheep grazing (Arnalds and Barkarson 2003); they are not important for agriculture in general as their use is often more linked to traditions rather than economic or sustainable land use.

2.4 Forestry in Iceland

At the time of Settlement, forest and shrubland covered up to 25 % of Iceland. At the beginning of the twentieth century, only three stands of a few hectares remained of the tall growing native birch (*Betula pubescens*), at Vaglaskógur, North Iceland, in Hallormsstaður, East Iceland, and Bæjarstaðarskógur, South Iceland (Fig. 2.10). These stands served

Fig. 2.9 Haymaking in West Iceland. Drying the hay was a major challenge during rainy summers until recently, but modern haymaking techniques are not as dependent on dry weather conditions. *Photo* Askell Thorisson, AUI

Fig. 2.10 An old native birch forest, Bæjarstaðarskógur, in Southeast Iceland. This is part of the few stands of tall growing birch that survived medieval times. *Photo* Asa L. Aradottir

as a seed source for extension of the birch forests throughout the twentieth century, but the spread was aided by natural succession in many areas protected from sheepgrazing (Aradottir and Eysteinsson 2005; Aradottir 2007). Forests now cover about 1.2 % of the country (Snorrason et al. 2005; Traustason and Snorrason 2008).

There has been considerable effort to increase the extent of forests in Iceland since the early twentieth century, both with natural and introduced species. The aims of the afforestation programs vary, but include aesthetics, nature, and/or soil conservation, but the aim of some of the plantations is to create timber forests (Aradottir et al. 2013). These forests are becoming a prominent part of the Icelandic landscape, especially in East Iceland. Part of this program is driven by aims to sequester carbon to balance emissions in relation to the UN Convention on Climate Change (see Sigurdsson et al. 2007). Restoring native birch forests still remains a priority (Aradottir and Eysteinsson 2005). Four introduced tree species are most common: Siberian larch (*Larix sibrica*), Sitka spruce (*Picea sitchensis*), lodgepole pine (*Pinus contorta*), and black cottonwood (*Populus trichocarpa*). A number of other tree species have been successfully introduced in Iceland. Currently, there is a debate about the use of these and other alien species in Iceland (see Chap. 4).

Other land use components that are increasing the stress on Icelandic soil systems include road construction and urbanization (Fig. 2.11). This has led to fragmentation of valuable ecosystems such as wetlands (Wald 2012), which still continues.

Fig. 2.11 Urbanization—road construction destroying valuable agricultural soil

References

Regarding punctuation and Icelandic characters in citations: see section on punctuation in the Preface

Adalsteinsson S (1981) Origin and conservation of farm animal populations in Iceland. Zeitschrift für Tierzüchtungsbiologie 98:255–264

Aradottir AL (2007) Restoration of birch and willow woodland on eroded areas. In: Halldorsson G, Oddsdottir ES, Eggertsson O (eds) Effects of afforestation on ecosystems, landscape and rural development. TemaNord 508:67–74

Aradottir AL, Eysteinsson Th (2005) Restoration of birch woodlands in Iceland. In: Stanturf JA, Madsen P (eds) Restoration of boreal and temperate forests. CRC Press, Boca Raton, pp 195–209

Aradottir AL, Petursdottir T, Halldorsson G, Svavarsdottir K, Arnalds O (2013) Drivers of ecological restoration: Lessons from a century of restoration in Iceland. Ecol Soc 18:33

Arnalds O, Barkarson BH (2003) Soil erosion and land use policy in Iceland in relation to sheep grazing and government subsidies. Environ Sci Policy 6:105–113

Arnalds O, Thorarinsdottir EF, Metusalemsson S, Jonsson A, Gretarsson E, Arnason A (1997) Jarðvegsrof á Íslandi (soil erosion in Iceland). The Icelandic Soil Conservation Service and the Agricultural Research University (now Agricultural University of Iceland), Reykjavik. http://jardvegsstofa.lbhi.is/lisalib/getfile.aspx?itemid=1806

Arnalds O, Thorarinsdottir EF, Metusalemsson S, Jonsson A, Gretarsson E, Arnason A (2001) Soil erosion in Iceland. The Soil Conservation Service (Landgræðsla ríkisins) and the Agricultural Research Institute (Rala), Reykjavík. http://jardvegsstofa.lbhi.is/lisalib/getfile.aspx?itemid=1807. (Translation of the original 1997 publication)

Björnsson H, Jonsson T, Gylfadottir SS, Olason EO (2007a) Mapping the annual cycle of temperature in Iceland. Meteorologische Zeitschrift 16(1):45–56

Björnsson H, Olason EO, Jonsson T, Henkirksen S (2007b) Analysis of a smooth seasonal cycle with daily resolution and degree day maps for Iceland. Meteorologische Zeitschrift 16:57–69

CAFF (Conservation of Arctic Flora and Fauna) (2001) Arctic flora and fauna: status and conservation. Edita, Helsinki

Crochet P, Johannesson T (2011) A data set of gridded daily temperature in Iceland, 1949–2010. Jökull—Icel J Earth Sci 61:1–17

Crochet P, Johannesson T, Jonsson T, Sigurdsson O, Björnsson H, Palsson F, Barstad I (2007) Estimating the spatial distribution of precipitation in Iceland using a linear model of orographic precipitation. J Hydrometeorol 8:1285–1306

Crofts R (2011) Healing the land. Soil Conservation Service of Iceland, Gunnarsholt. ISBN 978-9979-9295-5-0

Einarsson MA (1984) Climate of Iceland. In: Van Loon (ed) World survey of climatology, vol 15. Elsevier, Amsterdam, pp 673–697

Gudmundsson J, Brink SH, Gisladottir FO (2013) Preparation of a LULUCF land-use map for Iceland: development of the grassland layer and subcategories. Grassl Sci Eur 18:105–107

Helgadottir A, Eythorsdottir E, Johannesson T (2013) Agriculture in Iceland—a grassland based production. Grassl Sci Eur 18:30–43

Karlsson G (2000) Iceland's 1100 years. The History of a Marginal Society. Hurst & Company, London

OECD (2001) Environmental Performance reviews, Iceland. OECD, Paris

Olafsson H, Furger M, Brümmer B (2007) The weather and climate of Iceland. Meteorologische Zeitschrift 16:5–8

Saemundsson Th, Arnalds O, Kneisel C, Jonsson HP, Decaulne A (2012) The Orravatnsrustir palsa site in Central Iceland—Palsas in an aeolian sedimentation environment. Geomorphology 167–168:13–20

Sigurdsson BD, Snorrason A, Kjartanson BTh, Jonsson JA (2007) Total area of planted forests in Iceland and their carbon stocks and fluxes. In: Halldorsson G, Oddsdottir ES, Eggertsson O (eds) Effects of afforestation on ecosystems, landscape and rural development, Nordic Council of Ministers, Copenhagen, pp 211–217

Snorrason A, Kjartansson BT, Gunnarsson E, Eysteinsson T (2005) Global forest resources assessment update 2005. Iceland. Country report. FAO, Rome

Thorhallsdottir ThE (1997) Tundra ecosystems of Iceland. In: Wielgolaski FE (ed) Polar and alpine tundra, ecosystems of the world 3. Elsevier, Amsterdam, pp 85–96

Traustason B, Snorrason A (2008) Spatial distribution of forests and woodlands in Iceland in accordance with the CORINE land cover classification. Icel Agric Sci 21:39–47

Wald EC (2012) Land-use development in South Iceland 1900–2010. MSc thesis, Faculty of Biology and Environmental Science, University of Iceland, Reykjavík

Geology

3

3.1 Introduction

The volcanic geology of Iceland is a main factor shaping Icelandic soils, which will be given considerable attention in later chapters. Volcanism also explains much of Icelandic landscapes, together with the action of glaciers and frost. This chapter draws considerable material from a special issue of the *Icelandic Journal of Earth Sciences* titled *The Dynamic Geology of Iceland*, which was aimed to give an up-to-date overview of the geology of Iceland. It was published in 2008 (Issue 58). The *Icelandic Journal of Earth Sciences* is an ISI journal easily found in international databases, but has also the Icelandic name *Jökull* (meaning glacier in Icelandic). This journal has published a wealth of material on the various features of geology of Iceland. Ari Trausti Gudmundsson has published several books in English outlining the geology of Iceland, such as *Living Earth, the Outline of the Geology of Iceland* (Gudmundsson 2007), but his Icelandic 'Íslenskar eldstödvar' (Icelandic volcanoes) from 2001 is an excellent overview of the active volcanism on the island. A classical textbook on geology in Icelandic by Einarsson (1999) contains many descriptions and explanations of Icelandic geology, as would be expected. The volcanism and earthquakes continuously threatens the human population (Gudmundsson et al. 2008), which is explained in detail in a recently published book in Icelandic on natural hazards, a book that also provides a comprehensive and up-to-date review of the volcanic geology of Iceland, in Icelandic (Solnes et al. 2013).

3.2 Why Does Iceland Exist? The Mantle Plume Under Iceland

There is no doubt that the most important aspect of the geology of Iceland is its volcanism; it is a recent and active volcanic island. It is located on the Mid-Atlantic Ridge that separates the Eurasian and North American tectonic plates. These plates are drifting apart at a rate of about 2 cm each year (Hreinsdottir et al. 2001; Arnadottir et al. 2008; Einarsson 2008). Most of the Mid-Atlantic Ridge lies at the bottom of the ocean (Fig. 3.1) as would be expected with heavy basaltic rocks, in contrast to lighter weight continental crusts. Most of the volcanic activity on the Earth (>94 %) is associated with the active tectonic belts (Simkin and Siebert 2000). There are important exceptions to this distribution, such as cores of volcanic activity caused by *mantle plumes* or *hotspots*, which are driven by long-lived (millions of years) isolated plumes transferring heat from the mantle (e.g., Bjarnason 2008).

Basaltic chemical composition is typical for oceanic rift zones with relatively "raw" magma rising from the mantle filling in the gap created by the separating plates. These are the conditions both north and south of Iceland. The existence of Iceland is somewhat of an anomaly prompting the question: Why does Iceland exist? Fortunately for Icelanders, there are areas of high intensity of volcanic activity which occurs over a relatively short time (still millions of years), presently or in the past, which have been given the term *Large Igneous Provinces* or LIPs. Iceland is definitely one of these LIP areas. The Icelandic LIP area owes its existence to the plume of hot magma rising up from the mantle; the *hotspot* under Iceland. Bjarnason (2008) provided a detailed overview of the theories on the Icelandic hotspot.

The mantle plumes (hotspots on the surface) are relatively stationary on the globe, but the plates move slowly on top, a few centimeters each year, which, with continuous volcanic activity over millions of years, can create island arcs such as those of the Hawaii islands. Iceland represents a rather unique situation; the hotspot underlies the tectonic plate boundaries, and an island of basaltic rocks is formed and maintained, as the hotspot prevents these heavy basaltic rocks to sink to the bottom—or rather, the basaltic crust formed in Iceland is unusually thick: Einarsson (2008) described Iceland as a 300 × 500 km platform at the plate boundary and "on top of a hotspot presumed to be fed by a deep mantle plume." It should be noted here that the volcanic activity in Iceland is not only basaltic, but also

O. Arnalds, *The Soils of Iceland*,
World Soils Book Series, DOI 10.1007/978-94-017-9621-7_3
© Springer Science+Business Media Dordrecht 2015

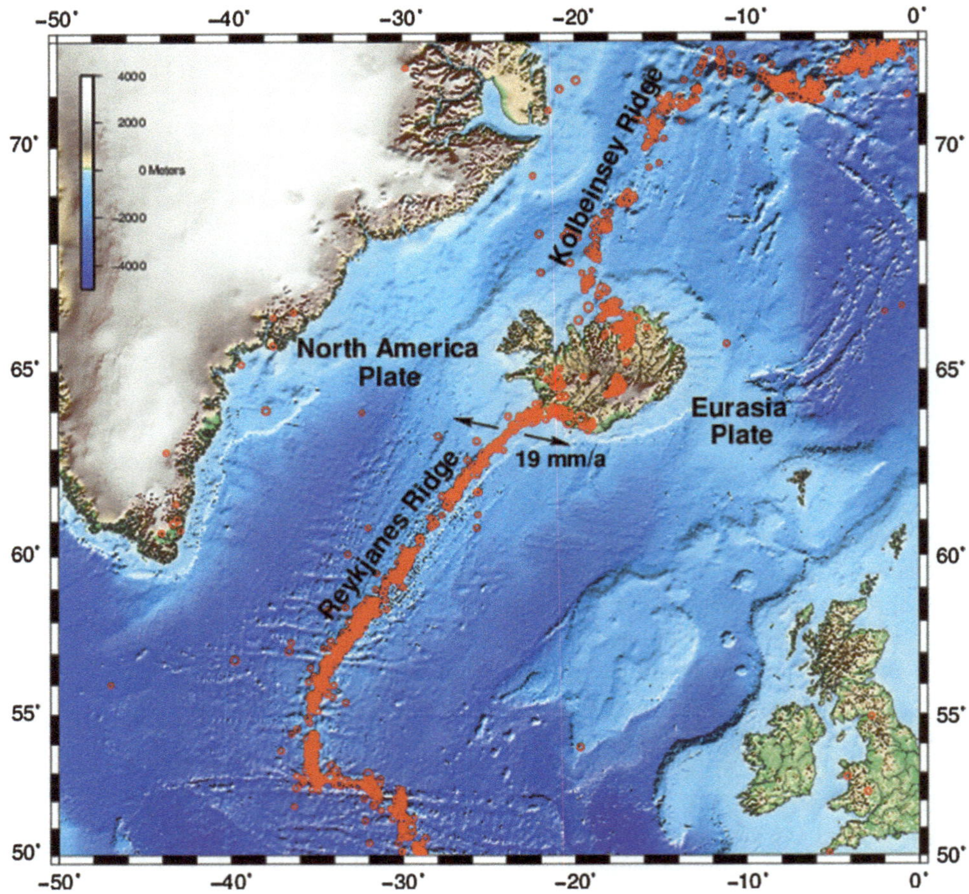

Fig. 3.1 Epicenters in Iceland and along the Mid-Atlantic Ridge system 1964–2004. Data are from the catalogs of the International Seismological Centre. Bathymetry is from GEBCO. Modified from Einarsson (2008) and Solnes et al. (2013), used with permission

andesitic and acidic (silicious) under some circumstances. This is because the magma is often subject to various kinds of interaction with the existing crust, with partial melting and differentiation of the chemical components of the magma on the journey from the mantle.

Earthquakes are very common as a result of the plate movements and hotspot activity, typically numbering 10,000 to >30,000 each year (Jakobsdottir 2008), while very few of them are powerful enough to cause damage.

The volcanic production of magmatic materials in Iceland rates among the most intense found on Earth, or 5 to >8 km^3 per century on average (see Thordarson and Höskuldsson 2008), making it the most productive hotspot and accounting for a sizeable share of the total of about 200 km^3 of magma produced on Earth each century (Francis and Oppenheimer 2004). Volcanic eruptions occur every 3–5 years on average (Thordarson and Höskuldsson 2008).

Iceland is being thrust to the east and west by the plate boundary cutting through the island. However, this boundary is not clear-cut. Two volcanic zones enter the south shores of Iceland, one at Reykjanes, making up the Reykjanes volcanic belt (RVB) and the western volcanic zone (WVZ) and the other in the Vestmann-Islands/Katla vicinity, making the eastern volcanic zone (EVZ) (Fig. 3.2, abbreviations in Fig. 3.7). A separate belt makes up the Snæfellsnes volcanic belt in West Iceland (SVZ). A series of volcanic systems are aligned on these zones or belts, each having a number from 1 to 30 in Fig. 3.2. Each volcanic system has a center (solid lines), but magma from below can be carried belowground along the fault systems and enter the surface in eruptions quite far from the center of these systems. The EVZ has been the most active part of Icelandic volcanism during the Holocene. Some of the most voluminous and frequent eruptions occur on the zone between the Katla system (14) and the Torfajökull system (17) to the southwest, extending northeast to the Grímsvötn (19) and Bárðarbunga (18) systems, producing numerous tephra layers found in soils (Fig. 3.3). Large recent eruptions associated with these systems include the Vatnaöldur eruption (about 870 AD, ~5 km^3 tephra, Larsen et al. 2013), which created a nice

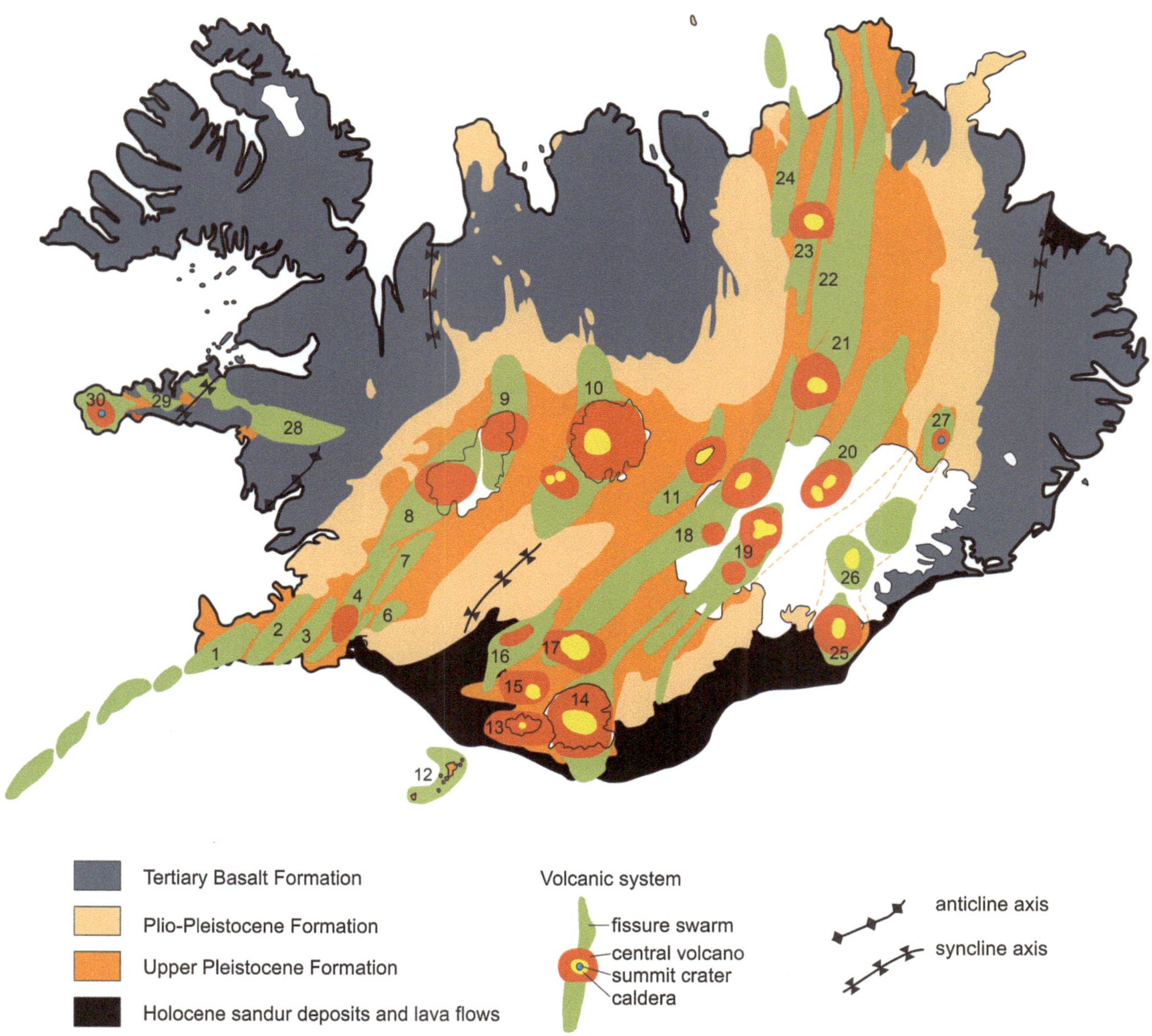

Fig. 3.2 Distribution of active volcanic systems among volcanic zones and belts in Iceland, based on Johannesson and Saemundsson (1998). Numbers indicate each volcanic system. The Katla (*14*), Hekla (*16*), Grímsvötn (*19*) and Bárðarbunga (*18*) systems are frequently referred to in the text. Reproduced with permission from Thordarson and Höskuldsson (2008)

marker for the Settlement in the soils, the gigantic Eldgjá eruption (934 AD, ~20 km^3, Fig. 3.3), Bárðarbunga-Veiðivötn eruption (1477 AD, ~15 km^3 tephra, Larsen et al. 2013), and the notorious Laki eruption in 1783 (>10 km^3 lava), which caused respiratory problems, famine, and deaths in Europe (see Thordarson and Höskuldsson 2008). The larger of these eruptions rate among the largest on Earth during historic times. Eruptions are also quite common in Mt. Hekla (Fig. 3.4) which has produced important tephra markers in Icelandic soils. The Askja system is frequently mentioned in the literature, and its activity includes an eruption that produced widespread rhyolitic ash in East Iceland in 1875.

3.3 Volcanoes and Active Volcanic Systems

The active volcanic systems shown in Fig. 3.2 vary considerably in nature. Many have well-defined centers with thermal activity and a fissure swarm extending in both directions, but not necessarily a defined volcano at the center, as is common at the Reykjanes ridge (RVB). Eruptions associated with such systems result in basaltic flows if they are in no contact with water or glaciers. Other systems have developed central volcanoes, with magma chambers underneath, such as Mt. Hekla (14), Eyjafjallajökull (13), Katla (14), and Öræfajökull (25). Various interactions and processes can occur in

Fig. 3.3 The volcanic landscape of the Southern Highlands. The crater in the foreground (the lake) is a part of the tens of kilometers long Eldgjá fissure (part of the Katla volcanic system) which formed in a huge volcanic eruption around 934 (about 20 km^3 of emitted materials; Thordarson and Höskuldsson 2008)

Fig. 3.4 Mt. Hekla, one of the most active volcanoes in Iceland. Restoration efforts to establish vegetation on land damaged by grazing and volcanic impacts in the foreground. *Photo* © Hreinn Oskarsson

Fig. 3.5 Glaciofluvial plain in front of Mýrdalsjökull glacier. The Katla volcanic system is located under the glacier to the left. Much of the Mýrdalssandur sandplain has formed during massive floods (jökulhlaups) in association with volcanic activity under the glacier. Glacial melt during summer also loads the plains, and the grayish part of the sandplain is among the most active dust sources in the country

relation to the magma chambers, which can result in andesitic and even silicious eruptions, which can be quite explosive (Mt. Hekla, Öræfajökull, Askja).

Many of the volcanic systems are capped with glaciers. When eruptions occur under glaciers, a violent mixing of hot lava and cold water occurs, creating explosive volcanism, which produces tephra (volcanic ash) where basaltic lava would be created without the glacial cap. This results in higher number of explosive eruptions than would otherwise be expected, from such glacially capped volcanoes as Katla (Fig. 3.5), Grímsvötn and Bárðarbunga, which are characterized by frequent eruptions. These volcanoes have produced large quantities of volcanic ash during the Holocene. The glacial meltwater formed during eruptions can result in extraordinary voluminous floods. The Katla floods are well known, reaching peak flows of 200,000–400,000 m^3 s^{-1} (Larsen 2000; Gudmundsson et al. 2008) which is of the same magnitude as the Amazon river spring floods. A flood after the Grímsvötn 1996 eruption reached about 45,000 m^3 s^{-1} discharge (Björnsson 2003) wiping out the road and bridges south of the glacier. These floods bring down extremely heavy sediment loads made of volcanic ash, and are similar in nature as lahars. They are termed *jökulhlaups* in Icelandic, a scientific term that has gained international acceptance for these natural events (see Rodolfo 2000). Jökulhaups have created extensive glaciofluvial floodplains in South Iceland, north of Vatnajökull and in various other areas (see Gudmundsson et al. 2008). They serve as active sources for aeolian materials that characterize the formation of Icelandic soils, as will be discussed later.

3.4 Tephra and Volcanic Ash

The airborne materials from volcanic eruptions are collectively termed *tephra* in geology. Tephra is the chief constituent of the parent materials of Icelandic soils, and it is therefore worthwhile to discuss tephra terminology briefly. The term *tephra* was introduced to the scientific literature by the Icelandic geologist Thorarinsson in 1944, "to describe all pyroclasts that leave a volcanic vent by air, regardless of

Fig. 3.6 A soil profile showing several tephra layers such as the Öræfajökull 1362 (AD) and the Settlement layer (about 870 AD). Numerous other tephra layers are evident, mostly from the Katla volcanic system. The soil is exceptionally thick and coarse, situated close to major aeolian sources and volcanoes. *Photo* © Bergrun Anna Oladottir

type, size, and shape" (see Larsen and Eiriksson 2008a). Tephra refers to materials that are primarily unconsolidated. The term *volcanic ash* is often used quite loosely about all tephra materials, but refers strictly speaking to tephra materials (or pyroclasts) that are less than 2 mm in diameter (Macdonald 1972; see also De Paepe and Stoops 2007; Arnalds 2013). "Pyroclasts" and "pyroclastic rocks" are also common terms, pyroclasts being a broader term than tephra, including both consolidated and unconsolidated materials (Schmid 1981; Manville et al. 2009), or for "all material ejected from volcanoes as solid fragments" (Francis and Oppenheimer 2004).

3.5 Tephrochronology

Thorarinsson (1944) introduced the term *tephrochronology*, for using tephra layers of known age for dating (see text by Steinthorsson (2012) about Sigurdur Thorarinsson). Tephrochronology has proved to be an extremely useful tool for a variety of research topics, as was recently reviewed by Lowe (2011). Tephrochronology has been applied in studies of pedogenesis, archeology, paleontology, geomorphology, dating and correlating volcanic events, and many more applications (e.g. Lowe 2011). An exceptionally large number of tephra layers are found in Icelandic soils (see Oladottir et al. 2012), where >150 eruptions have left markers in glacial ice and the Andosols over the past 1,140 years since settlement, giving many opportunities for practical applications (Larsen and Eiriksson 2008a, b). Sediments and soils at archeological sites can be dated with some accuracy, including the Norse settlement (Fig. 3.6), and the soils and pollen provide evidence for the climatological, ecological, and pedological history during the Holocene (Hallsdottir and Caseldine 2005; Gisladottir et al. 2011; Streeter et al. 2012).

3.6 Glaciation—The Quaternary

Most of Iceland was glaciated during the Quaternary, with pronounced effects on Icelandic landscapes. The glaciers began to expand from mountain massifs about 3.8 million years ago, but glaciation had become extensive about 2.5 million years ago (Eiriksson 2008). The glacier extended far out to what is now ocean around Iceland, coinciding with much lower sea levels due to water bound in the Quaternary glaciers. The glaciation was discontinuous, with each glacial event separated by ice-free conditions (Eiriksson 2008). During the glacial stage, deep valleys were cut into the Tertiary lava stacks, which are prominent features of the present day North, West, and Eastern Iceland. The glacier nearly obliterated extinct volcanic systems from the surface, but geothermal heat is still common within such areas, such as in the area around Reykjavík.

The consolidated volcanic tephra that make the rocks that formed under glaciers is termed hyaloclastite, but is also often referred to as *palagonite*, though the Icelandic term is *móberg*, constituting the Móberg Formation in Iceland, covering about 11,200 km^2 or 11 % of Iceland (see Jakobsson and

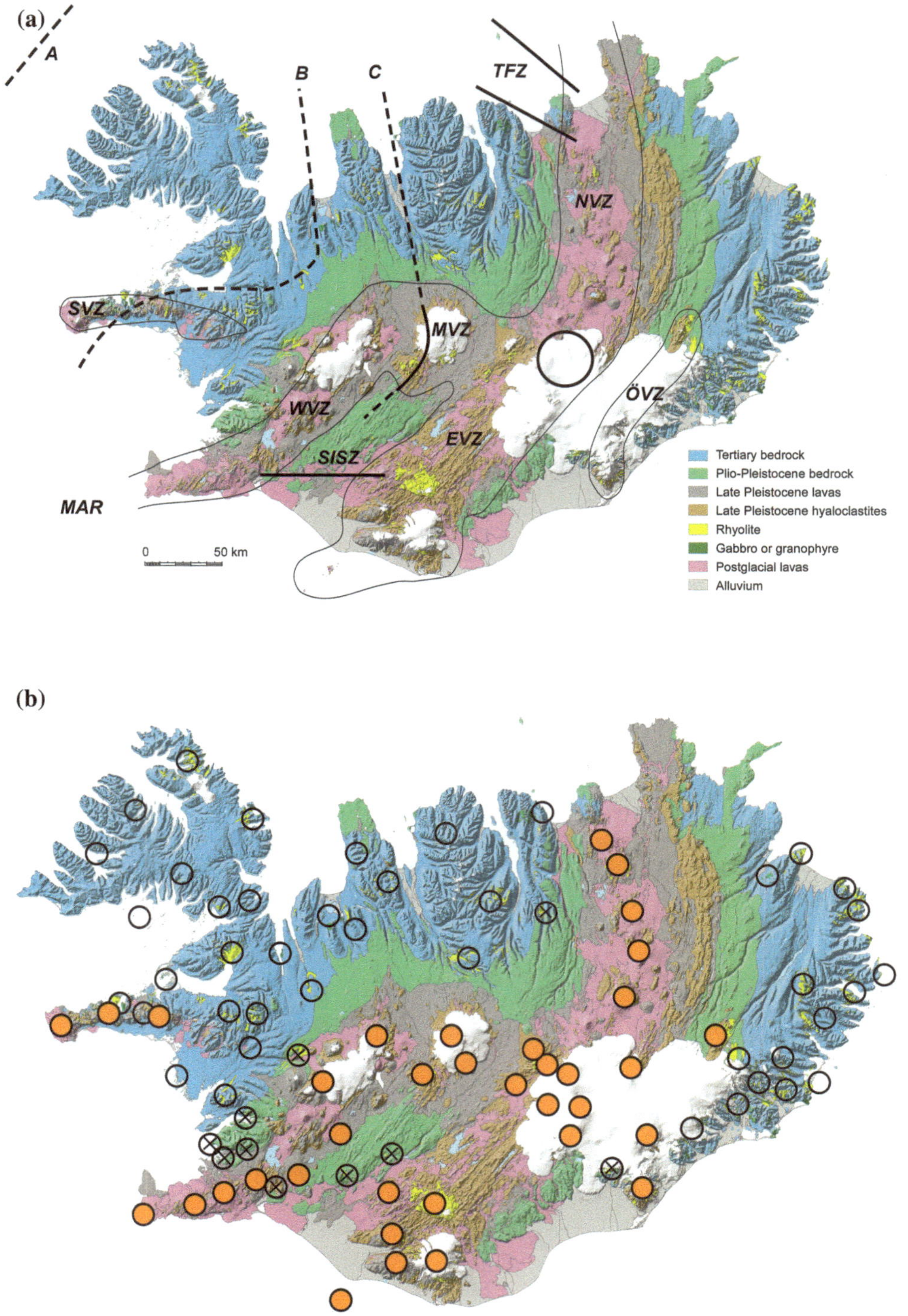

Fig. 3.7 **a** A simplified geologic map and the volcanic zones and rift systems in Iceland. **MAR** = Mid-Atlantic Ridge; **WVZ** = Western Volcanic Zone; **MVZ** = Mid-Iceland Volcanic Zone; **NVZ** = Northern Volcanic Zone; **EVZ** = Eastern Volcanic Zone; **SISZ** = South Iceland Seismic Zone; **TFZ** = Tjörnes Fracture Zone; **SVZ** = Snæfellsnes Volcanic Zone; **ÖVZ** = Öræfajökull Volcanic Zone. Circle represents the approximate center of the Iceland plume. Map modified from Johannesson and Saemundsson (1999); reproduced with permission from Hardarson et al. (2008). **b** Central volcanoes in Iceland. *Open circles* Tertiary (extinct) volcanoes; *crossed circles* Plio-Pleistocene volcanoes; *brown circles* active volcanoes. Based on Johannesson and Saemundsson (2003a, b); reproduced with permission from Hardarson et al. (2008)

Gudmundsson 2008; Fig. 3.7). Fissure eruptions under glaciers have formed ridges that are prominent in many highland areas in Iceland within the active volcanic zone, termed *móbergshryggir* in Icelandic (Fig. 3.8). Single event eruptions, which otherwise would have created shield volcanoes, have resulted in prominent table mountains or tuyas (*stapi* in Icelandic), but these mountains are a prominent part of the Icelandic landscapes, such as Eiríksjökull (West Iceland), and Herðubreið in the Northeast (Fig. 3.9).

The environmental conditions of the retreat of the glacier, with the onset of the Holocene, about 10,000 years ago was reviewed by Norddal et al. (2008, 2012).

Fig. 3.8 Palagonite ridges, unique and prominent part of the Icelandic volcanic landscapes. A view of the Fögrufjöll ridges (*center*), formed during fissure eruption under glacier during the last glaciation. Vatnajökull in background, Langisjór lake in the foreground

Fig. 3.9 Table mountain (stapi). Herðubreið is the national mountain of Iceland, located in the highlands of Northeast Iceland

3.7 Older Rocks—The Tertiary

With the Eurasian part of Iceland being thrust eastwards and the American plate pushed westwards, new rock materials are being produced within the active volcanic zone. Therefore rocks are gradually older moving away from the center of the island. The oldest rocks in Iceland are found in the far western and eastern parts of the country, dating several million years, but the oldest rocks are less than 20 million years old (Hardarson et al. 2008). These Tertiary rocks are of course quite young on a geological timescale.

The Tertiary rocks, (Fig. 3.7a) are predominately made of basalts with some andesites, as would be expected. They make up a thick stack of individual lava flows (see Hardarson et al. 2008). These individual lava flows vary considerably in thickness, from thin basalts (few meters) to some thicker andesitic and occasional silicic lavas (Fig. 3.10). The Tertiary regions cover about half of Iceland; in North and Northwest Iceland and along the eastern shore. Due to their relatively older age compared to the rocks of the volcanic belt, their porosity has been partially plugged by secondary minerals. The species composition of the minerals varies with the depth of burial (pressure) and age, but interesting and rare zeolites are among these minerals found in the Tertiary rocks. In certain locations, unusually pure calcite crystals are found, so-called *Iceland spar* (optical calcite), which was used extensively in Nicol prisms in microscopes to polarize light (Kristjansson 2007). As a result of plugging of the Tertiary rocks, the hydraulic conductivity is slower than of the various rocks of the active volcanic belt, which results in slower drainage and the common occurrence of wetlands within the Tertiary Formation (Arnalds and Oskarsson 2009).

The active tectonic belt is not stationary over a long time; it appears to have shifted periodically, which results in remnants of extinct volcanoes or volcanic systems within the Tertiary Formation (e.g., Sigmarsson et al. 2012; Hardarson et al. 2008). Activity within one system has characteristically formed a center with a magma chamber. In many areas, the Quaternary glacier has carved out valleys within these areas, exposing the roots or the interior of the systems, including the magma chambers (Fig. 3.11). The landscapes with exposed extinct central volcanoes are often quite colorful, with a range of rocks that have been altered by the intense heat associated with the volcanism and intrusive rocks such as granite, diorite, and gabbro. These areas tend to be beautiful and are particularly prominent along the fjords from the Southeast to Northeast Iceland (Fig. 3.12).

Many of the Tertiary valleys are quite deep, especially in Central North Iceland, with mountains characteristically >1,200 m above the valley bottoms (Fig. 3.13). Cirques glaciers are common at high altitudes, together with active rock glaciers and what is commonly perceived as old inactive rock glaciers at lower elevations (Gudmundsson 1995;

Fig. 3.10 Tertiary basalt stack in North Iceland. A degraded landscape, note vegetation remnants extending up the slopes

Fig. 3.11 An interior of an extinct volcano in Kjós, a part of the Skaftafell National Park. Colorful hydrothermally altered rocks are typical of such areas

see Chap. 10 on cryoturbation), although some of these features may be rock slides or rock avalanches, which are also very common (e.g., Jonsson et al. 2004).

3.8 The Magnificent Glaciers

Glaciers cover about 11 % of Iceland, but their effects on nature, soils in particular, are pronounced. They are a prominent part of the landscapes with the larger ones dominating the skyline of the interior and much of South–Southeast Iceland. Glaciers are termed 'jökull' in Icelandic (plural: jöklar). There is an excellent overview of Icelandic glaciers published in the journal *Jökull* (*The Icelandic Journal of Earth Sciences*) by Björnsson and Palsson (2008); also Helgi Björnsson, a prominent Icelandic glaciologist, published a comprehensive book on Icelandic glaciers in Iceland in 2009, however, in Icelandic.

There are four glaciers, usually termed the "large glaciers" (Fig. 3.14), which are Vatnajökull (8,100 km^2, Fig. 3.15), Hofsjökull (890 km^2), Langjökull (900 km^2), and Mýrdalsjökull (590 km^2, Fig. 3.16). Vatnajökull has an ice volume of 3,100 km^3, mean thickness of 380 m, and a maximum thickness >900 m (Björnsson and Palsson 2008). Other notable glaciers include Drangajökull (160 km^2) in the Westfjords, and Eyjafjallajökull (80 km^2) in the South, but there are numerous other glaciers larger than 10 km^2 in area in addition to smaller cirque glaciers and caps of single mountains.

Icelandic glaciers are sensitive to climatic fluctuations and are currently retreating, with Vatnajökull losing about 10 % (300 km^3) of its volume over the past century or so and the ice loss has accelerated over the past 15 years (see Björnsson and Palsson 2008). There are predictions that Iceland will lose the large glaciers in 150–200 years due to climate change (Björnsson and Palsson 2008).

Fig. 3.12 **a** Borgarfjörður Eystri area in East Iceland, characterized by colorful rocks associated with an extinct central volcano **b** Vestara-Horn in Southeast Iceland, with solid intrusive rocks formed in relation to Tertiary volcanism

The glaciers cap many active volcanoes, as was discussed earlier in this chapter. This results in the production of poorly crystallized volcanic glass during eruptions, which is deposited directly, often over wide ranging areas, during eruptions. However, glacial meltwaters, aided by glacial erosion, transport large quantities of these materials from underneath the glaciers to the glacial foreland. This sediment transport occurs both during regular summer melt of the glaciers and during jökulhlaups, catastrophic floods resulting from the volcanic eruptions or emptying out of subglacial reservoirs formed by glacial melt above active thermal areas (see Björnsson 2009). These materials tend to be very fine-grained and unstable, and are therefore subject to intense aeolian processes and redistribution of the volcanic materials, providing a major proportion of the parent materials for Icelandic soils (Fig. 3.17). Because of the importance of the aeolian materials, the aeolian environment is given consideration in a special chapter (Chap. 11).

Fig. 3.13 Valleys and fjords, carved out from the Tertiary Formation by the Pleistocene glacier (Westfjords)

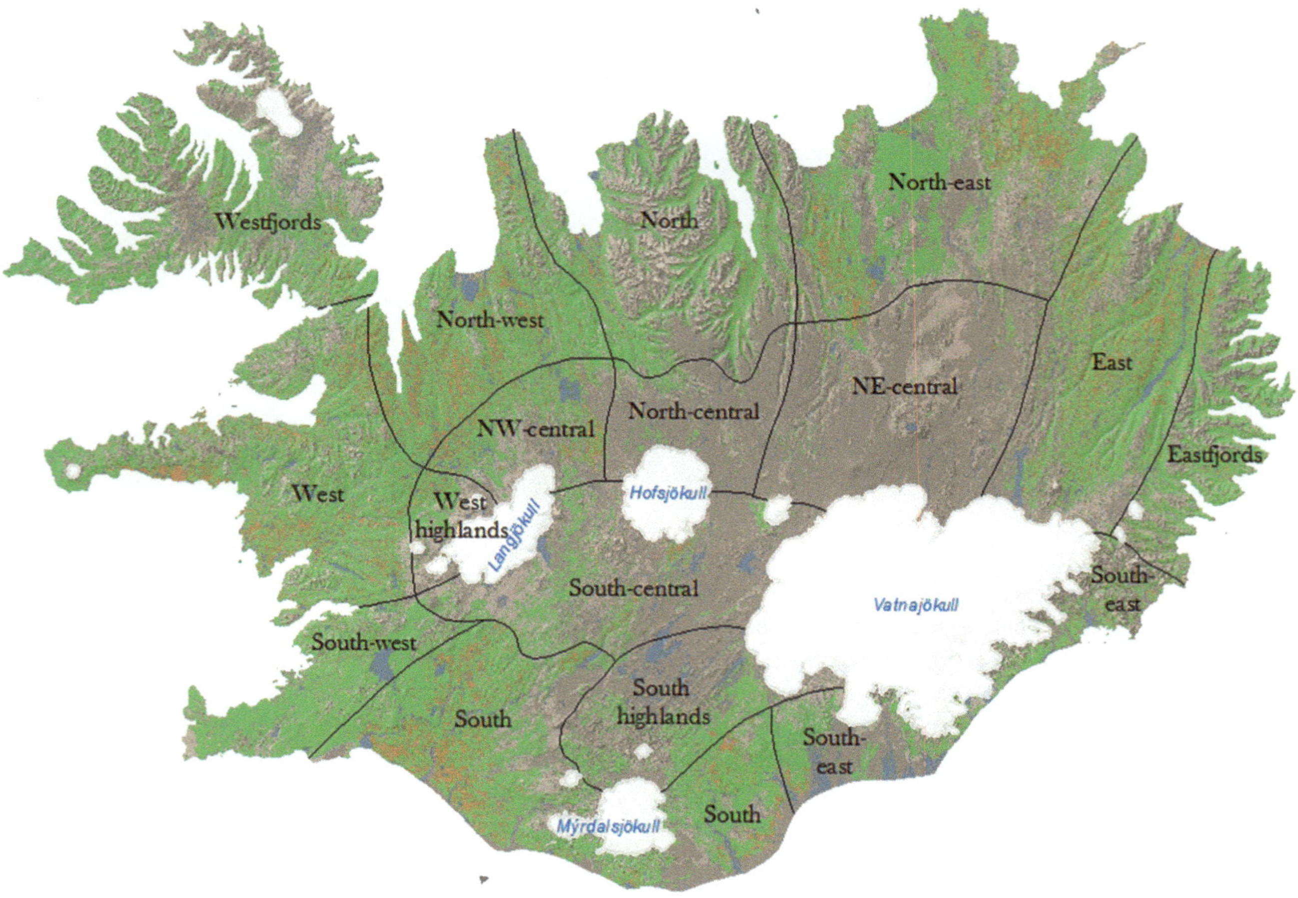

Fig. 3.14 The major glaciers and the division of Iceland into geographical areas referred to in the chapter

Fig. 3.15 Glacial landscape, Kverkfjöll in the northern part of Vatnajökull. This is an active volcanic system with thermal areas under the glacier, melting a lake in the glacier. The volcanic system is seen from a distance in Fig. 3.22. *Photo* © G.Kr. Johannesson

3.9 Rivers and Streams

A foreign traveler coming to Iceland may expect the volcanoes, the cold weather, and the glaciers to be the natural features that give the strongest impression. This is the experience of many, of course. But for others, this is not the case, but the abundance of water. The Icelandic landscapes are in part characterized by the numerous streams and larger rivers, waterfalls in the steep terrain, lake districts, wetlands, snow in winter, glaciers on the horizon, and, of course, rain.

There is a lot of precipitation that falls on Iceland, as can be expected for a mid-oceanic island (see Chap. 2). The fate of the water is quite variable, depending on the terrain, which includes such diverse surfaces as glaciers, sandy deserts, and vegetated land, with various soils and bedrocks; and slope angles from nearly level plains and steep slopes in mountainous areas. The average runoff rate is 1,460 mm year^{-1}, which is much higher than the average world runoff rate of 372 mm year^{-1} (Jonsdottir 2008).

There are not many publications that review Icelandic rivers with the exception of the book by Rist (1990) in Icelandic: *Vatns er þörf* ("Water is Needed"). The term "need" in the title refers in part to the need for water to generate hydropower energy. Most texts dealing with Icelandic rivers divide them into three main categories: (i) spring-fed rivers (groundwater fed rivers), often with relatively few water outlets making the river; (ii) surface runoff rivers ("dragár" in Icelandic), fed by system of ever smaller rivers and streams; and (iii) glacially fed rivers (Kjartansson 1945; Rist 1990; see also Jonsdottir 2008; Gislason 2008) (Figs. 3.18, 3.19, and 3.20). The Icelandic Meteorological Office operates elaborate systems to monitor water flow in most major Icelandic rivers, which is accessible in real time on the internet on the Icelandic Met Office website (www.vedur.is).

Most rivers, at least the larger ones, are a mixture of these three categories. Nonglacial rivers are usually a mixture of surface runoff and groundwater components, and the longer glacial rivers include both groundwater and surface runoff water (Fig. 3.19). There are notable exceptions, such as the

Fig. 3.16 Mýrdalsjökull glacier. The Katla volcanic system rests under the glacier

Fig. 3.17 The Skaftá river in South Iceland during a windy day. Sediments are blown from the floodplain toward the southern shore

Fig. 3.18 Brúará river. Nearly purely spring-fed river in South Iceland, most of the water appearing in a canyon below an extensive desert area around Hlöðufell table mountain. The water flow tends to be quite stable, but sometimes surface runoff during snow melt events increases the water flow

Fig. 3.19 Jökulsá á Fjöllum glacial river in Northeast Iceland during flooding caused by the summer melt of Vatnajökull glacier. There is a substantial spring water component in the river, which is dwarfed during peak summer flow from the glacier

Fig. 3.20 Surface runoff river in West Iceland. The water level is quite variable, the river being low during dry spells, but flooding during high intensity rains and snow–thaw events

short but powerful glacial rivers along the south and southeast coast which are purely glacially fed. Many spring-fed rivers within the active volcanic belt receive limited surface runoff due to high infiltration rates and permeability of the bedrock (except during thaw events on frozen ground). Surface runoff characterizes the areas outside the volcanic zone, such as the valleys of the Westfjords, West, Northwest, North, and Eastern Iceland and in parts of the South, with varying amounts of spring water. There is some glacial input from the numerous cirque glaciers in most major rivers in North Iceland.

There is a notable difference between the spring fed and the surface runoff rivers in that the spring-fed rivers have quite stable water flow over the year. The surface runoff rivers can have extremely fluctuating flow, often low in winter and during dry spells, but typically >10 times the average flow during spring floods and sudden thaw events in winter (Fig. 3.21). Here, the soil cover on the watershed has a dominant influence on the stability of the water flow. As the main water discharge in glacially fed rivers occurs during summers, when the glaciers melt and return much of the yearly snow accumulation, hydroelectrical power generation relies on reservoirs to accumulate the water and regulate the water flow over the year (Fig. 3.22). These reservoirs have large impact on Icelandic landscapes, some drowning valuable fully vegetated ecosystems, while the energy generated does not involve burning of fossil fuels. The reservoirs have resulted in rather stable water flow in some of the major glacial rivers, especially the river Þjórsá in the South, Blanda in the Northwest, and Lagarfljót in the East. The electricity generation with large reservoirs remains a subject of heated environmental debates in Iceland.

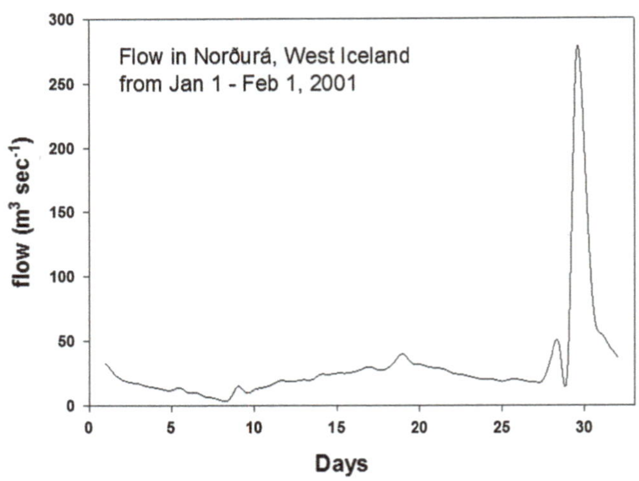

Fig. 3.21 Flow in the surface water fed Norðurá in West Iceland prior and during one rain and snow melt event in January. The water flow increases from about 3 to 270 m^3 s^{-1}, but is fast to recede again. Draining of wetlands and poor state of some of the rangelands within the watershed (low water infiltration in winter) amplifies the effect of such rain/thaw events. Data: Icelandic Met Office

Fig. 3.22 The Hálslón Reservoir in East Iceland, the Kverkfjöll volcanic system in the background. The water table fluctuates 40–70 m each year, being low in spring, with a potential of large-scale wind erosion of lake sediments over the surroundings. Most of the energy generated from reservoirs of this nature is used for aluminum smelters and other heavy industries

References

Regarding punctuation and Icelandic characters in citations: See section on punctuation in the Preface

Arnalds O (2013) The influence of volcanic tephra (ash) on ecosystems. Adv Agron 121:331–380

Arnalds O, Oskarsson H (2009) Soil map of Iceland (Íslenskt jarðvegskort). Náttúrufræðingurinn 78:107–121 (English summary, table and figure legends)

Arnadottir T, Geirsson H, Jiang W (2008) Crustal deformation in Iceland: plate spreading and earthquake deformation. Jökull—Icel J Earth Sci 58:59–74

Bjarnason ITh (2008) An Icelandic hotspot saga. Jökull—Icel J Earth Sci 58:3–16

Björnsson H (2003) Subglacial lakes and jökulhlaups in Iceland. Glob Planet Change 35:255–271

Björnsson H (2009) Icelandic glaciers (Jöklar á Íslandi). Opna, Reykjavík

Björnsson H, Palsson F (2008) Icelandic glaciers. Jökull—Icel J Earth Sci 58:365–368

De Paepe P, Stoops G (2007) A classification of tephra in volcanic soils. A tool for scientists. In: Arnalds O, Bartoli F, Buurman P, Oskarsson H, Stoops G, Garcia-Rodeja E (eds) Soils of volcanic regions of Europe. Springer, Heidelberg, pp 119–125

Einarsson P (2008) Plate boundaries, rifts and transforms in Iceland. Jökull—Icel J Earth Sci 58:35–58

Einarsson Th (1999) "Geology" (Jarðfræði, myndun og mótun lands). Mál og Menning, Reykjavik

Eiríksson J (2008) Glaciation events in the Pliocene-Pleistocene volcanic succession of Iceland. Jökull—Icel J Earth Sci 58:315–329

Francis P, Oppenheimer C (2004) Volcanoes, 2nd edn. Oxford University Press Inc., New York

Gisladottir G, Erlendsson E, Lal R (2011) Soil evidence for historical human-induced land degradation in West Iceland. Appl Geochem 26:S28–S31

Gislason SR (2008) Weathering in Iceland. Jökull—Icel J Earth Sci 58:387–408

Gudmundsson A (1995) Rock falls or rock glaciers? (Berghlaup eða urðarjöklar?) Náttúrufræðingurinn 64:177–186
Gudmundsson AT (2001) Icelandic volcanoes (Íslenskar eldstöðvar). Vaka-Helgafell, Reykjavik. ISBN 9979-2-1596-8 (in Icelandic)
Gudmundsson AT (2007) Living Earth. Outline of the geology of Iceland, Mál og Menning/Edda, Reykjavik. ISBN 978-9979-3-2777-6
Gudmundsson MT, Larsen G, Höskuldsson A, Gylfason AG (2008) Volcanic hazards in Iceland. Jökull—Icel J Earth Sci 58:251–268
Hallsdottir M, Caseldine C (2005). The Holocene vegetation history of Iceland, state-of-the-art and future research. In: Caseldine C, Russell A, Hardardottir J, Knudsen O (eds) Iceland—modern processes and past environments. Developments in quaternary science, vol 5. Elsevier, Amsterdam, pp 319–334
Hardarson BS, Fitton JG, Hjartarson A (2008) Tertiary volcanism in Iceland. Jökull (Icel J Earth Sci) 58:161–178
Hreinsdottir S, Einarsson P, Sigmundsson F (2001) Crustal deformation at the oblique spreading Reykjanes Peninsula, SW Iceland: GPS measurements from 1993–1998. J Geophys Res 106:13803–13816
Jakobsdottir S (2008) Seismicity in Iceland. Jökull—Icel J Earth Sci 58:75–100
Jakobsson SP, Gudmundsson MT (2008) Subglacial and intraglacial volcanic formations in Iceland. Jökull—Icel J Earth Sci 58:179–196
Johannesson H, Saemundsson K (1998) Geological map of Iceland, 1:500,000. Bedrock geology. Icelandic Institute of Natural History and Iceland Geodetic Survey, Reykjavik
Johannesson H, Saemundsson K (1999) Geological map of Iceland, 1:1,000,000. Icelandic Institute of Natural History, Reykjavik
Johannesson H, Saemundsson K (2003a) Extinct central volcanoes of Iceland. Geological map. Iceland GeoSurvey, Reykjavik
Johannesson H, Saemundsson K (2003b) Active central volcanoes of Iceland. Geological map. Iceland GeoSurvey, Reykjavik
Jonsdottir JF (2008) A runoff map based on numerically simulated precipitation and a projection of future runoff in Iceland. Hydrol Sci 53:100–111
Jonsson HB, Petursson HG, Norddahl H (2004) Was it a Rock Avalanche the formed Vatnsdalshólar in northern Iceland? Náttúrufræðingurinn 72:129–138 (in Icelandic, extended English Summary)
Kjartansson G (1945) Íslenskar vatnsfallategundir (classification of Icelandic streams). Náttúrufræðingurinn 15:113–126 (in Icelandic)
Kristjansson L (2007) On some applications of Iceland spar prisms in biological sciences and medicine Náttúrufræðingurinn 76:37–48 (in Icelandic, English summary)
Larsen G (2000) Holocene eruptions witin the Katla volcanic system, south Iceland. Characteristics and environmental impact. Jökull—Icel J Earth Sci 49:1–28
Larsen G, Eiriksson J (2008a) Holocene tephra archives and tephrochronology in Iceland—a brief overview. Jökull—Icel J Earth Sci 58:229–250
Larsen G, Eiriksson J (2008b) Late Quaternary terrestrial tephrochronology of Iceland—frequency of explosive eruptions, type and volume of tephra deposits. J Quat Sci 23:109–120
Larsen G, Gudmundsson MT, Einarsson P, Thordarson Th (2013) Bárðarbunga. In: Solnes J, Sigmundsson F, Bessason B (eds) Natural hazards in Iceland: volcanic and seismic hazard (in Icelandic). University of Iceland Press, Reykjavík, pp 253–261
Lowe DJ (2011) Tephrochronology and its application: a review. Quat Geochronol 6:107–153
Macdonald GA (1972) Volcanoes. Prentice-Hall Inc., Englewood Cliffs
Manville V, Nemeth K, Kano K (2009) Source to sink: a review of three decades of progress in the understanding of volcaniclastic processes, deposits, and hazards. Sediment Geol 220:136–161
Norddahl H, Ingolfsson O, Petursson HG, Hallsdottir M (2008) Late Wechselian and Holocene environmental history of Iceland. Jökull—Icel J Earth Sci 58:343–364
Norddahl H, Ingolfsson O, Petursson HG (2012) Deglaciation of Iceland and relative sea-level changes (Ísaldarlok á Íslandi). Náttúrufræðingurinn 82:73–86 (in Icelandic, extended English summary)
Oladottir BA, Larsen G, Sigmarsson O (2012) Deciphering eruption history and magmatic processes from tephra in Iceland. Jökull—Icel J Earth Sci 62:21–38
Rist S (1990) Water is Needed (Vatns er þörf). Bókaútgáfa Menningarsjóðs, Reykjavík
Rodolfo KS (2000) The hazards from lahars and jökulhlaups. In: Sigurdsson H, Houghton B, McNutt SR, Rymer H, Stix J (eds) Encyclopedia of volcanoes. Academic Press, New York, pp 957–971
Schmid R (1981) Descriptive nomenclature and classification of pyroclastic deposits and fragments: recommendations of the IUGS Subcommission on the Systematics of Igneous Rocks. Geology 9:41–43
Sigmarsson O, Martin E, Paquette J-L, Bosse V, Geirsson K (2012) Gliðnunarhraði Íslands metinn með aldursgreiningum á megineldstöðvum Austurlands (constant spreading rate through the geological history of Iceland). Natturufraedingurinn 82:105–111 (in Icelandic, English summary)
Simkin T, Siebert S (2000) Earth's volcanoes and eruptions: An overview. In: Sigurdsson H, Houghton B, McNutt SR, Rymer H, Stix J (eds) Encyclopedia of volcanoes. Academic Press, New York, pp 249–261
Solnes J, Sigmundsson F, Bessason B (eds) (2013) Natural hazards in Iceland: volcanic and seismic hazard (in Icelandic). University of Iceland Press, Reykjavík
Steinthorsson S (2012) Sigurður Þórarinsson (1912–1983). Jökull—Icel J Earth Sci 62:3–20
Streeter R, Dugmore AJ, Vesteinsson O (2012) Plague and landscape resilience in premodern Iceland. PNAS 109:3664–3669
Thorarinsson S (1944) Tefrokronologiska studier pa Island. Geogr Ann 26:1–217
Thordarson T, Höskuldsson A (2008) Postglacial volcanism in Iceland. Jökull—Icel J Earth Sci 58:197–228

4 Vegetation and Ecosystems

4.1 Introduction

The aim of this chapter is to provide a general overview of the vegetation and terrestrial ecosystems in Iceland. Soil types associated with each vegetation class are discussed briefly, but the classification system used for the soils is explained in Chap. 6, which deals with the Icelandic classification scheme.

There is no comprehensive overview of ecosystems in Iceland available in English. Hördur Kristinsson published a popular flora, available in both English and Icelandic (Kristinsson 2010), and the Icelandic Institute of Natural History maintains a web page with accessible information about botany, geology, zoology, and biotypes (www.ni.is). The Agricultural University of Iceland (formerly, Agricultural Research Institute) and the Icelandic Institute of Natural History have continuously been mapping the vegetation at various scales and for diverse purposes. Products include vegetation maps in 1:40,000 and 1:25,000 scale for a part of Iceland, and the Institute for Natural History has simplified this information in a map in 1:500,000 (Gudjonsson and Gislason 1998). The Agricultural University of Iceland has produced a vegetation classification map with relatively good resolution (1:25,000), a part of the so-called *Nytjaland* database (AUI Icelandic Farmland Database). This database is used extensively for this chapter. The AUI Farmland Database is also the underlying information for the most extensive land cover classes in the European CORINE dataset for Iceland. Information presented in this chapter, in addition to the AUI Farmland Database, is also drawn from various publications and maps from the Icelandic Institute of Natural History and other sources, such as Einarsson (2005) and Einarsson and Arnalds (2011).

4.2 Vegetation Classes and Common Plant Species

The flora of Iceland does not consist of many native species of vascular plants, which counts fewer than 500 species. In addition, there is a range of lichens (about 750 species) and mosses (about 600 species) and cyanobacter (Icelandic Institute of Natural History webpage www.ni.is; www.floraislands.is). The national flower of Iceland is *Dryas octapetala*, which is common throughout Iceland.

The AUI Farmland Database land-cover (Nytjaland) was created based on supervised classification of satellite images (www.lbhi.is/vefsjá; Arnalds et al. 2003; Arnalds and Barkarson 2003; Gisladottir et al. 2014). It uses 10 classes for vegetation in addition to glaciers/snow and open water. A map of Iceland drawn from the database is presented in Fig. 4.1. The aerial extent of each class, divided between elevation classes is presented in Table 4.1. Following is a short description of each class, based on the original Icelandic text (Arnalds et al. 2003). For more detailed account of the vegetation species, see Kristinsson (2010) and a summary for the highlands by Thorhallsdottir (1997).

Cultivated land and hayfields (1,723 km^2). Dairy cows and sheep are kept indoors during winter; hay-production for winter feeding can therefore be considered as the backbone of agricultural activities in Iceland (see Helgadottir et al. 2013). The soil types under cultivated land vary considerably, from various Brown Andosols and Gleyic Andosols, Histic Andosols to Histosols.

Shrubs and forests (1,205 km^2). This category includes land dominated by willow (*Salix phylicifolia*, *S. arctica* and *S. lanata*) and mountain birch (*Betula pubescens*) shrubs over about 50 cm height (Fig. 4.2). Understory consists most

O. Arnalds, *The Soils of Iceland*,
World Soils Book Series, DOI 10.1007/978-94-017-9621-7_4
© Springer Science+Business Media Dordrecht 2015

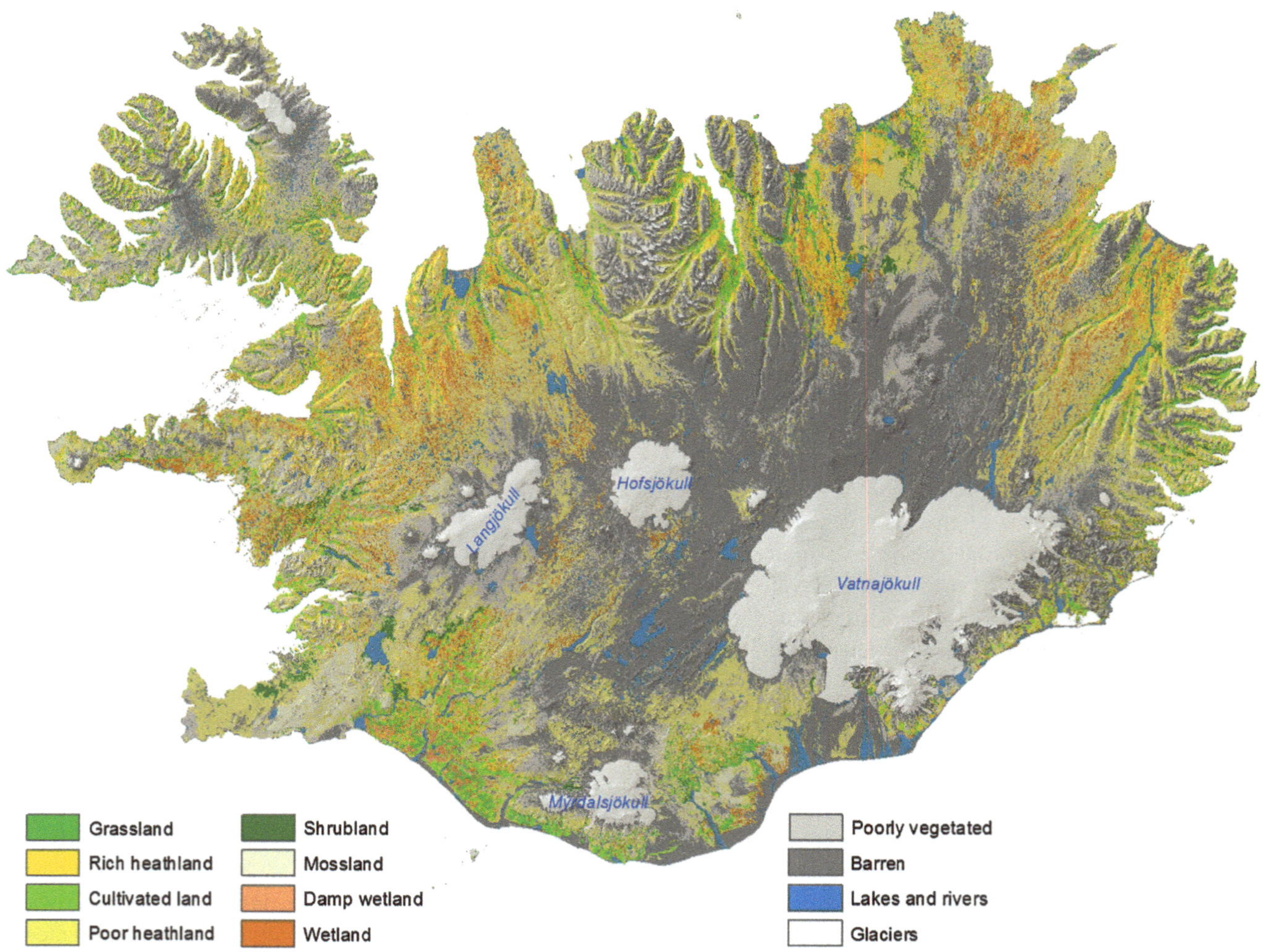

Fig. 4.1 A map of vegetation classes in Iceland. AUI Farmland Database classes (Nytjaland). The wetlands are shown as *pinkish-orange* color, but deserts (poorly vegetated and barren land) in *gray-tones*. Poor heathland has the largest extent of the vegetated classes. Prepared by Sigmundur Helgi Brink. AUI/SHB/OA. © AUI

often of forbs, grasses, and heath vegetation. Birch ecosystems were dominant ecosystems in Iceland prior to the Settlement (Aradottir and Arnalds 2001; Aradottir et al. 2001), as will be discussed later in the book. The primary soil types under forests are Brown Andosols, but also Histic Andosols to some extent in pristine and vigorous systems.

Grasslands (2,375 km^2). Grasslands occur where growing conditions are favorable, with ample soil moisture, commonly in depressions or in toe-slope positions, protected by snow in winter, but also in the lowland plains. Forbs are often prominent. Grasslands are also common on alluvial materials near streams. Grasslands include some former wetland areas that have been drained. Soil types under grasslands are quite variable (all are Andosols, however), but they are usually high in organic matter in surface horizons (>6 % C).

Saturated wetlands (3,968 km^2). The AUI Farmland Database (Nytjaland) separates between *saturated* wetlands and *damp* wetlands ('hálfdeigja'), the latter having aquic soils but partial drainage leads to a mixture of plants characteristic of wetlands and species more common within the dryland vegetation classes. Both these vegetation classes would have soil types qualifying for the aquic soil moisture regime. The wetland species include *Carex* spp., such as *Carex bigelowii, C. lyngbyei, C. rostrata, C. chordorrhizia, and Equisetum* spp. Cotton grass (*Eriophorum angustifolium*) is quite common as is heathland vegetation discussed below. The wetlands are usually fully vegetated unless they have been overgrazed by horses (which is common in some of the lowlands). The soils of saturated wetlands range from Gleyic Andosols within areas of high aeolian deposition, Histic Andosols to Histosols, far from aeolian sources. Example of wetlands is shown in (Fig. 4.3).

Damp Wetlands (1,829 km^2) are wetland areas with partial drainage, often at the margin of the saturated wetlands ('hálfdeigja' in Icelandic). Species include *Carex* spp. such as *Carex bigelowii, C. nigra,* and *C. vaginata, Equisetum*

Table 4.1 Vegetion classes divided between 200 m elevation intervals

Elevation intervals (m)								
Vegetation class	0–200	200–400	400–600	600–800	800–1,000	>1,000	Total	%
	km^2							
Hayfields/cropland	1,678	44	1	0	0	0	**1,723**	1.7
Shrubs and forests	979	215	10	1	0	0	**1,205**	1.2
Grassland	1,618	591	151	14	1	0	**2,375**	2.3
Saturated wetlands	1,540	1,168	1,017	241	2	0	**3,968**	3.9
Damp wetlands	1,164	474	163	28	0	0	**1,829**	1.8
Rich heathland	3,902	2,173	632	105	23	8	**6,843**	6.6
Poor heathland	6,770	6,548	7,623	3,479	393	34	**24,847**	24.1
Moss	939	1,392	842	202	8	1	**3,384**	3.3
Half vegetated	2,136	2,762	4,281	3,278	996	174	**13,627**	13.2
Barren	3,073	2,312	6,700	10,318	5,447	1,938	**29,788**	28.9
Streams and lakes	1,278	235	467	181	52	17	**2,230**	2.2
Glaciers and snow	72	195	272	875	1,445	8,242	**11,101**	10.8
Total	**25,149**	**18,110**	**22,158**	**18,722**	**8,369**	**10,414**	**102,922**	**100**
Vegetated minus moss	17,651	11,213	9,596	3,868	418	42	**42,788**	41.6
Moss	939	1,392	842	202	8	1	**3,384**	3.3
Desert (barren + ½ veget.)	5,210	5,074	1,0981	13,595	6,444	2,112	**43,416**	42.2
Wetlands (sat. + damp)	2,704	1,642	1,180	269	2	0	**5,797**	5.6

Total and % of Icealnd shown in the far right column. The lowest portion of the table is a summary of the upper portion. Based on the AUI Farmland Database, and the AUI IGLUD database. © AUI

Fig. 4.2 Shrublands in the southern part of the Westfjords (Barðaströnd). The shrubs are mostly birch (*Betula pubescens*) with occasional *Sorbus* trees

spp. such as *E. arvense* and *E. palustre*, and also *Juncus arcticus*. *Eriophorum angustifolium* is very common. Other species include grasses such as *Agrostis capillaris*, *Deschampsia caespitosa*, *Festuca richardsonii*, *Luzula multiflora,* and heath vegetation. Willow species (*Salix* spp.) and dwarf-birch (*B. nana*) are common in damp wetlands and sometimes birch (*B. pubescens*). Groundwater level is usually high, allowing for easy pumping of water to the freezing front, often resulting in unusually high hummocks, even >1 m high (see Chap. 10, Cryoturbation) as are found in the

Fig. 4.3 Example of wetlands. These wetlands are near Lake Mývatn in Northeast Iceland, in areas of large aeolian contributions. The soils are Gleyic Andosols with <12 % C in surface horizons. *Photo* © Asa L. Aradottir

northwestern highlands. The range of soils of damp wetlands is the same as for the saturated wetlands above.

Wetlands (damp and saturated) make up 3,968 + 1,829 = 5,797 km^2 or about 6 % of Iceland according to the Nytjaland database. In addition, there is a considerable area that now classifies as cultivated land and grasslands that are drained wetlands, making the total wetland area to about 8,000 km^2.

Rich heathland (6,843 km^2). The Icelandic heathlands have also been termed 'dwarf shrub heath' (Einarsson and Arnalds 2011). They are separated into two classes: rich heathland and poor heathland in the AUI Farmland Database (Nytjaland), partly reflecting the condition relative to grazing history. Rich heathland is dominated by dwarf heathland vegetation, such as dwarf-birch (*B. nana*), blueberries (*Vaccinium uliginosum, V. myrtillus*), crowberries (*Empetrum nigrum*), common heather (*Calluna vulgaris*) and *Arctostaphylos uva-ursi*, but also willow species (*Salix phylicifolia*, *S. arctica* and *S. lanata*). Moss species are also common, such as *Rachometrium* spp. Most of the heath species are not or less sought after for grazing by sheep compared to forbs and grasses, but rich heathland also has a significant component of herbaceous plants that are good for grazing, both grasses and forbs. This separates the rich heathland from the poor heathland below. Brown Andosols are the main soil type under the rich heathland but some are Histic Andosols (Fig. 4.4).

Poor heathland (24,847 km^2). Poor heathland is by far the most extensive vegetated class, covering about one-fourth of the country. This class is dominated by the heath species and has usually a large component of moss, such as *Rachomitrium lanuginosum*. Good grazing plants are not as abundant as in the rich heathland (<10 %) and sometimes nearly absent, most often as a result of long-lasting continuous grazing. One can say that the poor heathland has

Fig. 4.4 Example of the rich heathland vegetation class in an area that was protected from grazing about 20 years ago. Flowers and grasses are increasing their dominance, but willow (*Salix phylicifolia*) and dwarf-birch (*Betula nana*) are abundant. The soils are Brown Andosols

largely been shaped by the grazing history. Poor heathlands are characterized by Brown Andosols, but also Gleyic Andosols to some degree (Fig. 4.5).

Moss (3,384 km^2). One of the distinctive characteristics of Icelandic vegetation is the abundance of surfaces covered by moss. Mosses also appear with prominent abundance in poor heathlands, often as a result of grazing (plants sought after for grazing decrease in cover with time), but also as a primary succession vegetation on recent lavas (Cutler et al. 2008; Magnusdottir and Aradottir 2011) such as in the Laki lavas in South–Southeast Iceland from 1783 (Fig. 4.6). This vegetation class is dominated by moss species, but it should be noted that mosses appear in most other vegetation classes. Common moss species include *Rachomitrium lanuginosum*,

Fig. 4.5 Poor heathland is the most common vegetation class. It is dominated by heath species, moss, and sometimes lichens (*white* on *photo*). Erosion spots are also common as seen here. The soils are Brown Andosols. *Photo* Fanney Osk Gisladottir/AUI

Fig. 4.6 Moss. This moss cover has established on the lava from the gigantic Laki eruption in 1783. The *photo* was taken a year after the area was disrupted by volcanic ash from the 2011 Grímsvötn eruption. The soils are very shallow (Leptosols and shallow Brown Andosols)

R. ericoides, Hylocomium splendens, and *Rhytidiadelphus squarrosus* (see Magnusdottir and Aradottir 2011; Icelandic Institute of Natural History webpage www.ni.is; www.floraislands.is). The land covered with moss has quite low production of green biomass and has limited value for grazing. Soil types include Brown Andosols and Leptosols and Vitrisols (vitric soils of the deserts, Vitric Andosols under WRB, Vitricryands under Soil Taxonomy see Sect. 6.5).

Poorly vegetated—half vegetated land (13,627 km^2). A large proportion of Iceland is poorly vegetated, but many of the desert like areas have scattered vegetation cover (15–50 %), often isolated hummocks of plants well adapted to these conditions such as *Cerastium alpinum, Silene acaulis, Armeria maritime, Thymus praecox* subsp. *arcticus* (Fig. 4.7). Willow and heath are also common within the vegetation patches or hummocks. The poorly vegetated

Fig. 4.7 Poorly vegetated land has more vegetation cover than barren land (*below*), but these two classes combined are considered desert land. The vegetation (15–50 %) is often characterized by mosses and plants adapted to desert conditions with extensive root systems to exploit limited water and nutrient resources. The soils are Cambic Vitrisols (Vitric Andosols under WRB) with patches of Brown Andosols (also Vitric Andosols under WRB). *Photo* Fanney Osk Gisladottir/AUI

Fig. 4.8 Barren land, a desert, in South Iceland. This area receives >1,200 mm of annual rainfall. Barren state of the land is maintained by the instability of the sandy surface, sand abrasion, water shortage (limited infiltration in winter, rapid evaporation in summer), and grazing. The soils are Sandy (Arenic) Vitrisols, (Vitric Andosols under WRB)

ecosystem is limited in carbon and nitrogen stocks and is dysfunctional in terms of nutrient and water cycling (see text on the desert environment). Poorly vegetated land occurs within desertified areas, but also at high elevations and areas disturbed by volcanic events and flooding. It occurs also within degraded areas with shallow soils, resulting in rock outcrops where part of the soils have been lost, such as in West Iceland.

This vegetation class is counted with the barren land (below) as desert land, which makes up 13,627 + 29,788 = 43,416 km^2 in total, or about 42 % of Iceland. These areas are referred to as 'sparsely vegetated land' by the Icelandic Institute of Natural History. The main soil types are various Vitrisols (vitric soils of deserts; Vitric Andosols under WRB, Vitricryands under Soil Taxonomy), but Leptosols also occur.

Barren land (29,788 km^2) is characterized by the lack of vegetation (Fig. 4.8). As the poorly vegetated land described above, the barren land is a dysfunctional ecosystem, with intense cryoturbation processes, limited infiltration during winter, rapid evaporation during summer, limited water storage, and lack of functional water and nutrient cycles (see Chap. 12). However, scattered vegetation occurs, often 5–15 % cover with the same plants as within the poorly vegetated class. The deserts (poorly vegetated and barren) surfaces consist of a variety of geologic surfaces, such as sand surfaces, lag gravel, and lavas. The main soils are Vitrisols (vitric soils of the deserts) and Leptosols.

4.3 Vegetation Cover and Relation to Elevation

Iceland has about 45 % vegetation cover in total according to the Nytjaland database (Table 4.1), a number that includes the land with moss cover. This is the poorest vegetation cover in all of Europe. Deserts cover >40 %, but a substantial part of the deserts were vegetated at the time of Settlement (see Chap. 12 on land degradation).

Considering the northerly location of Iceland, cold climate is an important constraint to vegetation growth and ecosystem function. There is generally a rapid drop in heat units for plant growth with increased elevation. Figure 4.9 shows changes in vegetation cover with elevation, using 200 m elevation intervals. Total vegetation cover (minus

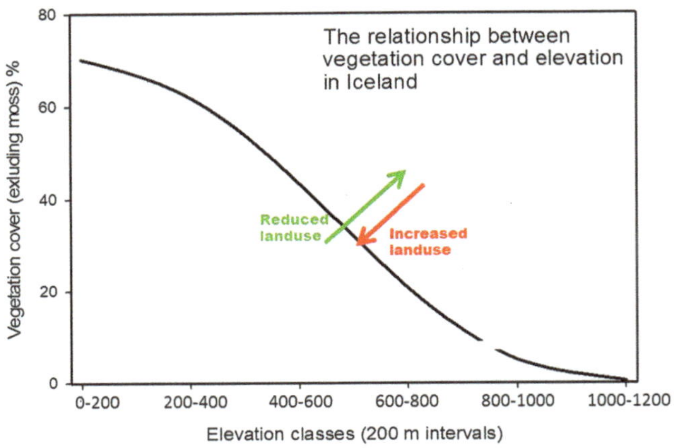

Fig. 4.9 Vegetation cover (%) as a function of elevation. *X-axis* represents 200 m elevation intervals. *Arrows* indicate a shift in the curve for vegetation cover; toward the *left* with land use (current situation) and to the *right* with reduced land use (and before Settlement). Data from the AUI Farmland Database (Nytjaland)

land with moss cover) is about 70 % for the 0–200 m elevation class and continues to be relatively high (62 %) with 200–400 m elevation but drops rapidly with height after that to 0.4 % above 1,000 m elevation (Fig. 4.9).

Grassland, shrubs/forests, and rich heathland are most prominent at the lower elevations, representing better growing conditions and resilience of the ecosystems in response to land use. The poor heathland becomes dominant vegetation class at higher elevations, together with desert landscapes.

With a gradual change from the forest/shrub types of vegetation, which dominated Iceland at the time of Settlement, to the current dominance of poor heathlands and deserts, there has been a dramatic shift in microclimate conditions. Shrubs allow for snow accumulation during winter, which isolates the soils and allows for increased infiltration of melt water. Surface wind-speeds are low within shrublands, but increase with reduced vegetation cover. Snow is blown off the barren surfaces, which really renders mean precipitation values meaningless for the desert surfaces: only a portion of the precipitation infiltrates into the soils. Furthermore, nonpermeable ice that forms in the desert soil blocks infiltration during winter (see Chap. 10 on frost and cryoturbation). Based on this, it can be postulated that only a third to half of the precipitation reaching the Icelandic deserts are leached into the soil. Much of rapid rainfall in summer is lost as gravitational water to the ground water as many of the desert soils are quite sandy with limited water holding capacity. Snow melt is lost as runoff due to impeded infiltration blocked by solid ice in the soil. The dark desert surfaces heat up in sunshine (often >50 °C), which leads to rapid evaporation in summer, sometimes aided by rapid hydraulic conductivity. From this, it is clear that water shortage is a fate of many of the deserts, in spite of ample rainfall; hence dry desert soil conditions can occur in moist climatic environments. These physical properties change rapidly when vegetation cover is re-established on the surface. The special conditions in Iceland have been used to shed a different light on concepts such as 'desert' and 'desertification', which, according to the UN Convention to Combat Desertification, is primarily confined to areas with low precipitation, while the ecosystem approach to such definition, considering the fate of the rain/soil moisture is a much more viable approach (Arnalds 2000). What matters is the fate of the water, not how much it rains.

4.4 Wetlands, Drainage and Agriculture

Wetlands rate among Iceland's most important ecosystems, with a key role in water and nutrient regulation and rich species diversity (Gunnarsson et al. 2006). Icelandic ecosystems support 21 internationally important populations of breeding bird species (Einarsson et al. 2002), and is responsible for a large proportion of the world population of some species (Wetlands International 2006). Iceland is an important stopover area for birds migrating across the Atlantic Ocean on their way between breeding habitats and wintering areas (Einarsson et al. 2002; see also Gunnarsson 2010; Gunnarsson et al. 2006).

Wetlands are 5,800 to about 8,000 km^2 in total depending on how they are defined. They occur in all geographic regions of Iceland and extend into highland elevations. However, about half of the wetlands are found below 200 m elevation. The most prominent wetlands occupy the southern lowlands, but wetlands are also common within the Tertiary Formation in West Iceland. Widespread wetlands are found in the northwestern and eastern highlands and they are an integral part of the deep valleys of the Tertiary Formation in Iceland.

Much of Icelandic wetlands have been drained for agricultural purposes. The main production on the drained areas is hay for winter, but grazing by cattle, sheep, and horses is an also important use of the drained wetlands. This drainage constitutes a widespread and pronounced ecosystem disturbance. The Icelandic government provided generous subsidies to enhance the drainage after World War II, in order to promote food safety in Iceland (Oskarsson 1998). This led to excessive drainage of land and a significant proportion of this area is not used for agriculture (see Fig. 4.10).

It is estimated that 3,400–3,900 km^2 of Icelandic wetlands have been drained, with about 32,000 km long ditches (Oskarsson 1998; AUI/IGLUD (Icelandic Geographic Land Use Database); Hallsdottir et al. 2013). Nearly 97 % of the wetlands in South Iceland lowlands have been drained (Thorhallsdottir et al. 1998) which gives an indication of the extent of disturbance. Drainage of wetlands is discussed further in Chap. 12.

Fig. 4.10 Drained wetlands in a valley bottom in the Loðmundarfjörður valley in the Eastfjords. Farming was abandoned a year after the drainage was put in. Such wetland disturbance has pronounced negative ecosystem impacts

4.5 The Desert Ecosystems

The deserts are perhaps the most conspicuous of all Icelandic geographic surfaces (Fig. 4.11). Desert areas are not common outside the dry regions of Earth, but most of the Icelandic deserts occur in areas of >500 mm of annual rain. Dark-colored or black volcaniclastic deserts are also rare on Earth, Iceland has by far the largest volcaniclastic desert surfaces (Edgett and Lancaster 1993; Arnalds et al. 2001). Their occurrence under a relatively humid oceanic climate makes them quite special, and also their vast extent, covering >40 % of Iceland (Table 4.1). The reason for the large extent of deserts is complex (see Arnalds 1987). Sometimes the explanation lies in destructive natural processes, such as volcanic activity and catastrophic flooding, especially in the

Fig. 4.11 Deserts are a prominent part of the Icelandic landscapes, but the surfaces include gravelly and sandy surfaces (shown). This area has very limited vegetation cover, which in part is due to the instability of the surface and continuous abrasion by the sand

Fig. 4.12 Biological soil crust in Iceland (*dark areas*). Various bryophytes are gaining foothold in the crust. A camera case (*lower left*) provides a scale

highlands and to some extent for the lowland floodplains. In other cases, land use is to blame, often in combination with natural events such as ash deposition or cold spells (see Arnalds 1987; Arnalds and Barkarson 2003; Aradottir and Arnalds 2001; Chap. 12). Continuous grazing in the communal grazing areas in combination with natural stresses are the main factors in many areas, including large areas at relatively low elevations (<500 m) in the southern lowlands, such as the communal areas south of Langjökull and Hofsjökull and in the neighborhood of Mt. Hekla and the Katla volcanic systems. One way of stating the impact of man on Icelandic ecosystems is that the curve in Fig. 4.9 has been shifted to the left and down as shown by the arrow in the figure. Further discussion of the desert types and erosion is provided in Chap. 12.

4.6 The Biological Soil Crusts

An important component of ecosystems (Fig. 4.12), however often overlooked due to its subtle appearance, is the *biological soil crust* or *biocrust* (see e.g., Belnap 2003; Johnson et al. 2012). Biological soil crusts consist of associations of cyanobacteria, algae, microfungi, lichens, bryophytes, and soil particles in different proportions within or immediately on top of the soil (Belnap 2001). It is usually only a few millimeter thick. Biological soil crusts are capable of stabilizing the soil surface to prevent erosion (Belnap 2003) and needle-ice formation (see Chap. 10 on Cryoturbation). It enhances soil aggregation, stabilizes the surface temperature, infiltration, reduces evaporation and winter runoff, and brings nitrogen into the ecosystem. Soil biological crusts can be used as an indicator of ecosystem health (Tongway and Hindley 1996; Bowker 2007). Biological crusts are extremely vulnerable to disturbance, with grazing disturbances being the most extensive (e.g., Belnap 2003; Gómez et al. 2012; Read et al. 2011), but burial by aeolian and/or volcanic processes is also detrimental (see Arnalds 2013). The biological soil crusts are often a necessary precursor to enhance primary succession in relation to restoration of severely degraded land in Iceland (e.g., Elmarsdottir et al. 2003). Furthermore, they are likely to be instrumental in building up adequate nitrogen levels for proper ecosystem functioning in the restored systems. They are therefore quite an important component of Icelandic ecosystems and ecosystem development.

4.7 Introduced and Invasive Species

The Icelandic ecosystems have evolved on an isolated island after the retreat of the Pleistocene glacier, without grazing animals, but with relatively few species of higher plants. These ecosystems are therefore particularly vulnerable to the introduction of new species into the existing species assemblages, especially species that are of invasive nature. von Schmalensee (2010a, b) has reviewed the influence of alien species in Iceland. She reported that several thousand species have been imported to Iceland since its Settlement, of which seven are particularly invasive: American mink (*Neovison vison*), Nootka lupine (*Lupinus nootkatensis*), cow parsley (*Anthriscus sylvestris*), Spanish slug (*Arion*

Fig. 4.13 Nootka lupine (Alaska lupine) is prominent on the Icelandic landscape in many places

lusitanicus), heath star-moss (*Campylopus introflexus*), white-tailed bumblebee (*Bombus lucorum*), and European physa (*Physella acuta*). There are several more species that are considered to pose a possible threat, including trees and shrubs such as rugosa rose (*Rosa rugosa*), lodgepole pine, (*Pinus contorta*), dark-leaved willow (*Salix myrsinifolia*), and the European rabbit (*Oryctolagus cuniculus*) (see von Schmalensee 2010b). Reindeer were introduced in East Iceland, but they are not considered an invasive species (von Schmalensee 2010b) and are an important part of the rural economy of East Iceland.

The Nootka lupine (Alaska lupine), which is a nitrogen fixing plant, was imported primarily for revegetation of the nutrient poor deserts (Fig. 4.13). This plant is economic in use and successful in early revegetation efforts, but its influence on long-term success of revegetation and ecosystem restoration can also be negative by preventing succession of native vegetation. It is spreading rapidly, mostly due to unrestricted seeding by the public and government organizations after 1980. Nootka lupine is increasingly becoming a prominent part of the Icelandic landscape, as are introduced tree species in some areas. The use of the Nootka lupine is currently debated and efforts are now being made to eliminate lupines (also the mink and the cow parsley) from some of the more sensitive nature protection areas; efforts that have had limited success to date.

References

Regarding punctuation and Icelandic characters in citations: See note on punctuation in the Preface

Aradottir AL, Arnalds O (2001) Ecosystem degradation and restoration of birch woodlands in Iceland. In: Wielgolaski FE (ed) Nordic mountain birch ecosystems. UNESCO/MAP book series, vol 27. Parthenon Publishing Group, London, pp 293–306

Aradottir AL, Thorsteinsson I, Sigurdsson S (2001) Distribution and characteristics of birch woodlands in North Iceland. In: Wielgolaski FE (ed) Nordic mountain birch ecosystems. UNESCO/MAP book series, vol 27. Parthenon Publishing Group, London, pp 51–61

Arnalds A (1987) Ecosystem disturbance in Iceland. Arct Alp Res 19:508–513

Arnalds O (2000) Desertification: an appeal for a broader perspective. In: Arnalds O, Archer S (eds) Rangeland desertification. Kluwer Academic Publishers, Dordrecht, pp 5–15

Arnalds O (2013) The influence of Volcanic tephra (ash) on ecosystems. Adv Agron 121:331–380

Arnalds O, Barkarson BH (2003) Soil erosion and land use policy in Iceland in relation to sheep grazing and government subsidies. Environ Sci Policy 6:105–113

Arnalds O, Gisladottir FO, Sigurjonsson H (2001) Sandy deserts of Iceland: an overview. J Arid Environ 47:359–371

Arnalds O, Thorsson J, Thorarinsdottir EF (2003) Landnýting og vistvæn framleiðsla sauðfjárafurða (Land use and ecofriendly sheep production). Rala report no 211. (Agricultural University of Iceland). ISSN 1010-0121

Belnap J (2001) Factors influencing nitrogen fixation and nitrogen release in biological soil crusts. In: Belnap J, Lange OL (eds) Biological soil crusts: structure, function, and management. Springer, Berlin, pp 241–261

Belnap J (2003) The world at your feet: desert biological soil crusts. Front Ecol Environ 1:181–189

Bowker MA (2007) Biological soil crust rehabilitation in theory and practice: an underexploited opportunity. Restor Ecol 15:13–23

Cutler NA, Belyea LR, Dugmore AJ (2008) The spatiotemporal dynamics of a primary succession. J Ecol 96:231–246

Edgett KS, Lancaster N (1993) Volcaniclastic aeolian dunes: terrestrial examples and application to Martian sands. J Arid Environ 25:271–297

Einarsson E (2005) Icelandic flora and vegetation (Flóra og gróður Íslands). In: Eldjarn O (ed) Íslandsatlas (Icelandic Atlas). Edda útgáfa, Reykjavík, pp 18–23 (in Icelandic)

Einarsson E, Arnalds O (2011) Vegetation and soils of the Thingvallavatn catchment. In: Jonasson P, Hersteinsson P (eds) Thingvallavatn—a unique world evolving. Opna Publishing, Reykjavik, pp 74–92

Einarsson O, Kristinsson H, Skarphedinsson KH, Ottosson JG (2002) Verndun tegunda og svæða: tillögur Náttúrufræðistofnunar Íslands vegna Náttúruverndaráætlunar (Nature protection suggestions). Náttúrufræðistofnun Íslands, Reykjavik

Elmarsdottir A, Aradottir AL, Trlica MJ (2003) Microsite availability and establishment of native species on degraded and reclaimed sites. J Appl Ecol 40:815–823

Gisladottir FO, Arnalds O, Brink SH (2014) Nytjaland (The Icelandic Farmland Database), Agricultural University of Iceland report no 49 (in Icelandic) (in prep.)

Gomez DA, Aranibar JN, Tabeni S, Villagra PE, Garibotti IA, Atencio A (2012) Biological soil curst recovery after long-term exclusion in the Monte Desert (Argentina). Changes in coverage, spatial distribution, and soil nitrogen. Acta Oecologica 38:33

Gudjonsson G, Gislason E (1998) Vegetation map of Iceland 1:500,000. General overview map. Icelandic Institute of Natural History, Reykjavik

Gunnarsson TG (2010) Votlendi og vaðfuglar í ljósi landnotkunar (Shorebirds and wetlands). Náttúrufræðingurinn 79:75–86 (English summary)

Gunnarsson TG, Gill JA, Appleton GF, Gislason H, Gardarsson A, Watkinson AR, Sutherland WJ (2006) Large-scale habitat associations of birds in lowland Iceland: implications for conservation. Biol Conserv 128:265–275

Hallsdottir BS, Wöll C, Gudmundsson J, Snorrason A, Thorsson J (2013) Emissions of greenhouse gases in Iceland from 1990 to 2011. National inventory report 2013. Environmental Agency of Iceland, Reykjavik

Helgadottir A, Eythorsdottir E, Johannesson T (2013) Agriculture in Iceland—a grassland based production. Grassl Sci Eur 18:30–43

Johnson SL, Kuske CR, Carney TD, Housman DC, Gallegos-Graves LV, Belnap J (2012) Increased temperature and altered summer precipitation have differential effects on biological soil crusts in a dryland ecosystem. Glob Change Biol 18:2583–2593

Kristinsson H (2010) Íslenska plöntuhandbókin. Blómplöntur og byrkningar. (A guide to the flowering plants and ferns of Iceland). Mál og Menning. Reykjavik (available in Icelandic, English and German)

Magnusdottir M, Aradottir AL (2011) Leiðir til þess að fjölga hraungambra og öðrum mosategundum (propagation of *Racomitrium lanuginosum* and three other species). Náttúrufræðingurinn 81 (3–4):115–122 (English summary)

Oskarsson H (1998) Wetland drainage of the western lowlands (Framræsla votlendis á Vesturlandi). In: Olafsson JS (ed) Icelandic wetlands. Háskólaútgáfan (University Publishing), University of Iceland, Reykjavik, pp 121–129

Read CF, Duncan DH, Vesk P, Elith J (2011) Surprisingly fast recovery of biological crusts following livestock removal in southern Australia. J Veg Sci 22:905–916

Thorhallsdottir TE (1997) Tundra ecosystems of Iceland. In: Wielgolaski FE (ed) Polar and alpine tundra, ecosystems of the world 3. Elsevier, Amsterdam, pp 85–96

Thorhallsdottir TE, Thorsson J, Sigurdardottir S, Svavarsdottir K, Johannsson MH (1998) Wetland drainage in S-Iceland (Röskun votlendis á Suðurlandi). In: Olafsson JS (ed) Icelandic wetlands. Háskólaútgáfan (University Publishing), University of Iceland, Reykjavik, pp 131–142

Tongway TJ, Hindley N (1996) Manual for assessment of soil condition of tropical grasslands. CISRO, Canberra

von Schmalensee M (2010a) Ecosystems in peril, part one: a review on alien, invasive species. (Vágestir í vistkerfum. – fyrri hluti. Stiklað á stóru um framandi ágengar tegundir) Náttúrufræðingurinn 80 (1–2):15–26 (English summary)

von Schmalensee M (2010b) Ecosystems in peril, part two: alien and invasive species in Iceland. (Vágestir í vistkerfum – Seinni hluti. Framandi og ágengar tegundir á Íslandi). Náttúrufræðingurinn 80 (3–4):84–102 (English summary)

Wetlands International (2006) Waterbird population estimates, 4th edn. Wageningen, The Netherlands

5 Andosols—Soils of Volcanic Regions

5.1 Introduction

The primary influence on most soils of Iceland is the volcanic nature of the parent materials (Fig. 5.1). This leads to the formation of *Andosols*, a special soil group under the WRB soil classification system (IUSS Working Group WRB 2006) due to the unique soil properties that develop. These soils are termed *Andisols* (soil order) under the US Soil Taxonomy (Soil Survey Staff 1999, 2003). As Andosols dominate Icelandic landscapes, it is worthwhile considering some of their main characteristics before discussing soils of Iceland specifically. The following chapters (Chap. 6 on classification and Chaps. 7, 8, and 9 on soil properties) refer back to this current chapter about Andosols in general.

The term 'Andosol' is derived from the Japanese, 'an' meaning dark, and 'do' connoting soil (Fig. 5.2). Andosols occupy a limited extent (<1 %) of the Earth's land surface, but many such areas are densely populated (McDaniel et al. 2011).

The properties of Andosols have been widely studied, but overview publications include a book by Shoji et al. (1993) and three special issues of scientific journals (Fernandez Caldas and Yaalon 1985; Bartoli et al. 2003; Arnalds and Stahr 2004). Overview chapters include works by Wada (1985), Kimble et al. (2000), Arnalds (2008), McDaniel et al. (2012), a monograph by Dahlgren et al. (2004), and an early compilation of benchmark papers on Andosols (Tan 1984). An overview of various aspects of soils of volcanic regions in Europe has also been published (Arnalds et al. 2007) as a result of joint European scientific cooperation (COST project), where Icelandic soil resources are among the studied soils.

Andosols are uncommon in northern Europe and in the circumpolar regions as can be seen on the map in Fig. 5.3. Other large Andosol areas at northerly latitudes include areas in Kamchatka and Alaska, following the so-called "Pacific Ring of Fire" volcanic activity. The map clearly shows the uniqueness of the Icelandic soil environment, which led to many misunderstandings in the past, where rhyolitic tephra layers were interpreted as eluvial horizons of Podzols, while others have placed some soil of Iceland in the Leptosol/Entisol category on the basis of lack of soil genesis (which is far from the truth, as will be explained in the subsequent chapters).

5.2 Classification

The development of the concept of Andosols has roots in the U.S. Soil Taxonomy, first presented as the Andept suborder of Inceptisols (Smith 1986), but from 1990 as Andisols, based on work of an international working group (ICOMAND) as was reviewed by Parfitt and Clayden (1991). The concept of the Andosol soil group, as used in the WRB, is similar to that of Soil Taxonomy (see Shoji et al. 1996). The central concept of Andosols is soil development of volcanic ejecta (Fig. 5.1). The tephra (mostly volcanic glass, see Chap. 3) weathers rapidly which results in the precipitation of so-called 'short-range order' minerals and/or 'metal–humus complexes', a process that is sometimes referred to as 'andozolization' (e.g., Duchaufour 1977). These colloidal constituents provide Andosols with characteristic properties such as low bulk density, high-organic content, rapid hydraulic conductivity and high water retention, variable charge characteristics, thixotropy, and strong phosphate retention. Andosols/Andisols are classified based on a measure of the colloidal constituents (ammonium oxalate extractable Al and Fe (Al_{ox} + ½Fe_{ox} > 2 %), low bulk density (<0.9), and high phosphate retention (>85 %). These criteria have been harmonized between the WRB and Soil Taxonomy systems. The diagnostic properties reflect the product of soil genesis. A unique feature of Andosols is that they can contain up to 25 % carbon (both systems), but are still considered as Andosols if other criteria are met. This is because the dominant influence of andic soil properties and because Andosols tend to accumulate large amounts of organic matter (see Shoji et al. 1993). However, Andosols (WRB) and Andisols (Soil Taxonomy) also include soils with a large component of

O. Arnalds, *The Soils of Iceland*,
World Soils Book Series, DOI 10.1007/978-94-017-9621-7_5
© Springer Science+Business Media Dordrecht 2015

Fig. 5.1 The eruption in Eyjafjallajökull in 2010 added new parent materials to soils of southern Iceland. The central concept of Andosols is soils that develop in volcanic ejecta. Several farms are seen in the foreground

non-weathered volcanic glass, by lowering the diagnostic limits if volcanic glass is prominent (need > 0.4 % Al_{ox} + ½Fe_{ox} and other requirements also lowered or waived). This is quite important in the context of soils of Iceland, as a substantial proportion of the country is barren with surfaces dominated by volcanic glass, consisting of soils primarily classified as Andisols or Andosols (Arnalds and Kimble 2001; Arnalds 2004; Arnalds and Oskarsson 2009).

5.3 The Colloidal Constituents of Andosols—The Soils of Iceland

5.3.1 Clay Mineral Formation in Andosols

The clay minerals that make Andosols so unique, together with the metal–humus complexes, are allophane, imogolite, ferrihydrite, and halloysite (see Wada 1989; Shoji et al. 1993; Dahlgren 1994; Harsh et al. 2002; McDaniel et al. 2011). These are not layered lattice clay minerals such as smectite and kaolinite, but are described by terms such as 'spherical', 'tubular', and 'gel-like'. Their crystallinity has been subject to debate and these constituents have been described as 'amorphous', 'X-ray amorphous', 'poorly crystalline', 'noncrystalline', and 'short-range order'.

The reason for the formation of the Andosol clays is linked to the nature of the tephra parent materials of Andosols, which weather rapidly, resulting in high concentrations of Al, Fe, and Si. The poorly crystalline (short-range order) morphological forms of these minerals are the result of rapid crystallization of Al and Si (allophane and imogolite) and Fe (ferrihydrite) from such soil solution. However, these minerals are not exclusive to Andosols as they are also commonly found in Podzols, but to a lesser degree. The weathering of basaltic tephra is rapid, resulting in areas of high chemical denudation (Stefansson and Gislason 2001),

Fig. 5.2 A dark colored Andosol from the southern lowlands. Note, clear granular structure in the A horizons and roots extending deep into the soil. Carbon rich horizons reach for the entire depth of the soil as is common with Andosols

and rapid formation of allophane and ferrihydrite. The surface area of basaltic tephra can be quite high or >10 m^2 g^{-1} (Wolff-Boenisch et al. 2004). Weathering rates of silicious (rhyolitic) tephra are slower than in basalt with less abundance of cations released to maintain the pH (but the tephra in Iceland is mainly basaltic). Intense weathering often results in more acidic soils, especially in humid-wet areas (Fig. 5.4). Dry climates can alternatively result in relatively unaltered parent materials, especially if the tephra is silicious.

5.3.2 Allophane and Other Andosol Clays: Odd 'Creatures' Among Clay Minerals

Clay minerals are fundamental in giving soils the properties that are essential for their function. They are extremely small, most of them made of sheets of two-dimensional layers, one on top of the other with space between the sheets. Therefore, they are termed phyllosilicates, layer silicates/minerals, or sheet silicates/minerals. The extremely small size of the layers, usually measured in Angstroms or nanometers, and the space between the layers results in tremendously large surface area, which can be as high as 200 m^2 for each single gram of soils! This results in the ability of the clays to hang on to water adsorbed to the surfaces of the clays for extended periods of time but the water is still available for plant growth, long after the last rains. Clays also retain cations that are important for plant growth, such as Ca^{++}, Mg^{++}, K^{+}, and Na^{+}, where roots of plants can access and utilize these ions. Common clay minerals include smectite, illite, kaolinite, vermiculite, goethite, and gibbsite, and many of these are used for various industrial purposes.

The major clay mineral formed in Andosols is allophane. It is very different from the conventional clay minerals as it is less crystallized and has a spherical shape. It still has enormous surface area and chemical reactivity, but allophanic soils lack the cohesion of other clayish soils. Its cation exchange capacity is pH dependent, in contrast to such minerals as the common smectite. Allophane has a tendency to stimulate carbon accumulation in soils, and this organic matter improves physical and chemical properties of the soils, as well as their fertility. Other common clay minerals in Andosols are imogolite, ferrihydrite, and halloysite. Ferrihydrite is especially common in Iceland.

Allophane is an aluminum and silica mineral that forms hollow spherules about 5 nm in diameter (see Fig. 5.5). The atomic ratio between Al and Si is somewhat variable, most commonly 1–2, but values <1 occur (Parfitt and Kimble 1989) and are common in Iceland (e.g., Arnalds and Kimble 2001). These minerals have an extremely large surface area and a charge that is pH dependent (variable charge), which increases rapidly with increasing pH. In addition, allophane has considerable anion exchange properties.

Imogolite is tubular and often appears thread-like viewed with a transmission electron microscope. It usually has an Al/Si ratio close to 2, and has similar properties as allophane (Fig. 5.6).

Ferrihydrite is a poorly ordered Fe(III) mineral (Schwertmann 1985), consisting of well-aggregated spherical particles (Bigham et al. 2002) which often appear with gel-like structure. Its structure has been debated and ideas about the nature of ferrihydrite are still evolving. Ferrihydrite is very common in Andosols, especially where the parent materials are rich in iron, as in Iceland. It has a large surface area and a pH-dependent cation and anion exchange capacity (Bigham et al. 2002).

Halloysite is a common mineral in Andosols, especially in Si-rich environments, and is often associated with dry

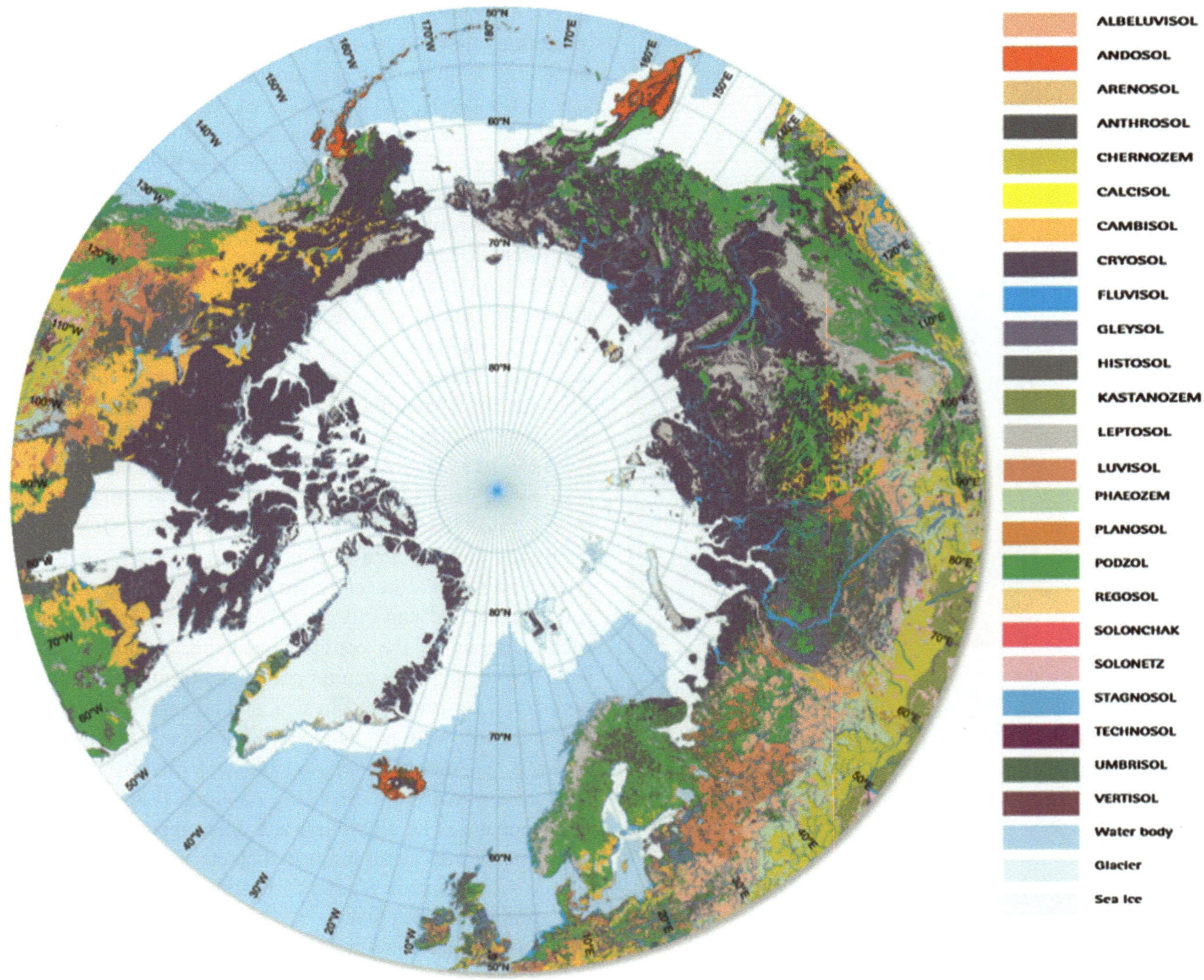

Fig. 5.3 The Circumpolar Soil Map from the Soil Atlas of the Northern Circumpolar Region (Jones et al. 2010). Andosols are *red* on the map, with the largest Andosol areas in Iceland, Kamchatka and Alaska

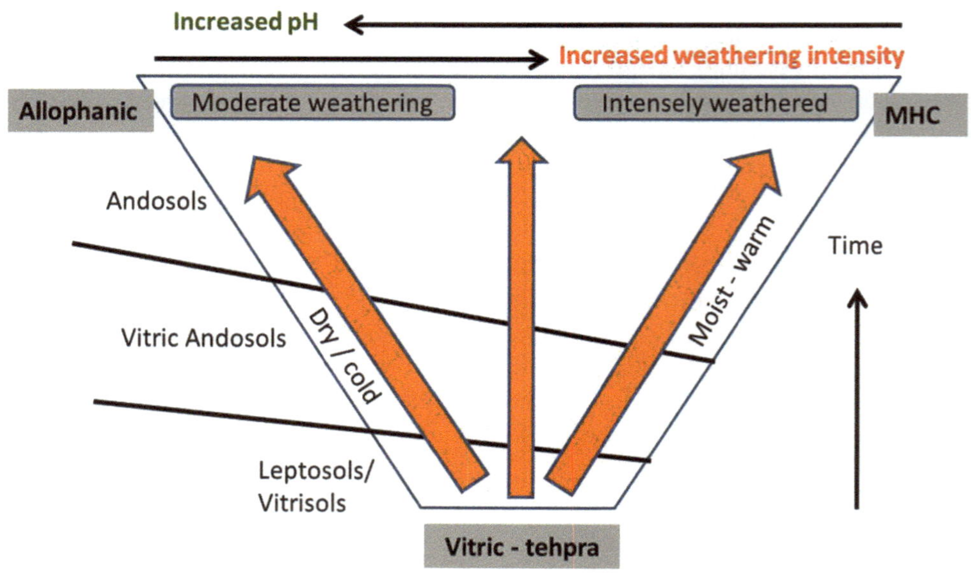

Fig. 5.4 The effect of weathering intensity on the formation of colloids in young Andosols. Metal humus complexes (MHC) are characteristic of highly intense weathering of tephra, while allophanic soils are more common in less weathered environments. Based on Arnalds (2013)

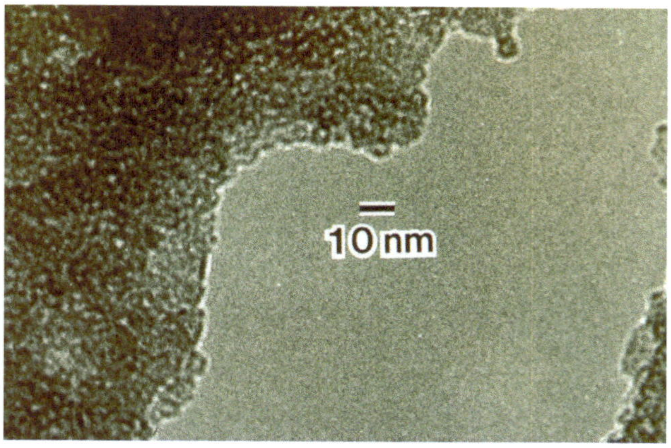

Fig. 5.5 A photo obtained with a transmission electron microscope (TEM) of allophane in an Icelandic soil. See also Wada et al. (1992)

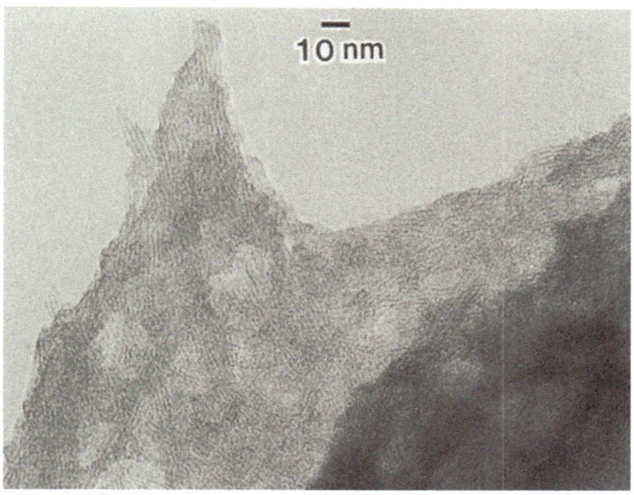

Fig. 5.6 A photo obtained with a transmission electron microscope (TEM) of imogolite in an Icelandic soil. See also Wada et al. (1992)

environments or a distinct dry season (Dahlgren et al. 2004; McDaniel et al. 2012). The morphology of halloysite is believed to be closely related to kaolinite (see White and Dixon 2002). Halloysite is often reported as representing more weathered environments than allophane dominated soils (e.g., Ndayiragije and Delvaux 2004). Other minerals are found in many Andosols, especially when Andosols become mature, with the Andosol minerals being transformed to other minerals such as kaolinite, smectite, Al/Fe oxides, and chloritized 2:1 minerals (e.g., Shoji et al. 1985). Opaline silica is also often reported in young Andosols, especially under grassland vegetation (e.g., Shoji et al. 1993).

5.4 Allophane–Humus and Metal–Humus Complexes

The original concept of Andosols ('an-do') reflects the dark color of many Andosols, which mainly results from the accumulation of organic matter. Large contents of organic matter characterize well-developed Andosols. Appreciable amounts of carbon are found at depths, and the distribution is often quite erratic. This is typical of many soils in Iceland, with high carbon levels deep in the soils. There are two main pathways of organic accumulation in Andosols: the formation of allophane-organic matter complexes and metalhumus complexes.

Allophane and organic matter form bonds that are relatively stable, which results in soils that commonly have >6 % C in both A and B horizons. Another mode of carbon accumulation is when Al^{3+} and Fe^{3+} form stable bonds with organic matter by ligand exchange (metal–humus complexes). This means carbon accumulation is effective at a relatively low pH, while allophane-humus accumulation is enhanced at higher pH. Research has confirmed the stability of organic constituents in Andosols, which can be >100,000 years old in Hawaii (Torn et al. 1997).

In some areas, other environmental factors can enhance the accumulation of organic materials in Andosols, such as poor drainage and cold climate, resulting in OC of 12–20 %, which is the case in Iceland (Arnalds 2004).

5.5 Andosols and the Carbon Cycle

Volcanoes emit large quantities of greenhouse gases into the atmosphere. However, the soils that form in the volcanic deposits, Andosols, have a tendency to accumulate organic materials as was previously described. Andosols store more carbon reserves per unit area than other dryland soils, often >30 kg C m^{-2} (Batjes 1996; Eswaran et al. 1993). In addition, calcium released from the weathering of basaltic volcanic materials, such as found in Iceland, have a tendency to react with CO_2 from the atmosphere to form bicarbonate and finally $CaCO_3$ that precipitates both in rocks and in the ocean (see Gislason 2008). However, land degradation in relation to overexploitation of volcanic soils contributes to the release of greenhouse CO_2 by reducing the carbon levels of the soils (Shoji and Takahasi 2002). Therefore, restoration of degraded areas in volcanic regions can result in rapid sequestration of carbon from the atmosphere (Arnalds et al. 2013). Zehetner (2010) concluded that carbon accumulation in volcanic soils did not offset CO_2 releases from volcanoes, but

Fig. 5.7 Thick volcanic tephra deposits in the vicinity of Mt. Hekla. This stack has remained relatively unweathered for thousands of years and contains both andesitic and rhyolitic tephra. Thick deposits of this kind commonly form consolidated hardpans in Iceland and elsewhere. They are called 'móhella' in Icelandic. *Photo* © Bergrun Anna Oladottir

he used lower carbon accumulation averages than experienced in Iceland (Arnalds et al. 2013, see also Chap. 9), and did not consider the $CaCO_3$ formation in soils, rocks, and oceans through weathering of Andosols.

5.6 The Three Axes of Andosols: Vitric, Allophanic, and Metal–Humus Complex Andosols

The parent materials made of tephra (Vitric or *vitrandic*), the higher pH allophanic (*sil-andic*) Andosols, and Andosols dominated by metal–humus complexes or organo-mineral complexes (*alu-andic*) can be viewed as the three 'end-members' of Andosols (Shoji et al. 1996). These axes are represented by the corners of the triangle in Fig. 5.4; they are important in the context of soils in Iceland as is discussed in the next chapter. The rate of cation release and the pH of the soil solution largely determine whether allophanic Andosols or soils dominated by metal–humus complexes are formed. Allophane formation is favored by a high pH, while it does not form when pH is under 5. Under such acidic pH conditions, the formation of metal–humus complexes becomes a dominant process. The 'alu-andic' soils often contain considerable amounts of phyllosilicates, such as chloritized 2:1 minerals (Shoji et al. 1985), under a variety of climatic conditions, which contribute to their physical and chemical behavior (see also Ndayiragije and Delvaux 2004).

The vitric materials are relatively unweathered volcanic deposits (Fig. 5.7), but the subsequent weathering rate is dependent on the chemical composition, climate, and biotic factors. Thickness of the tephra is also important, as thick deposits with low biological activity are more likely to remain less weathered than thin deposits spread onto functional ecosystems. The thicknesses of the tephra (ash layers) in soils in Iceland are quite variable, but ash deposition events are considerably more frequent than in most other volcanic areas, with numerous thin tephra layers (<1 mm) in most Andosol profiles. Few relatively thick layers (>10 mm) occur in close vicinity of the active volcanic systems.

5.7 Physical Properties

Many peculiar physical properties give Andosols unique characteristics, such as strong silt-sized aggregation and thixotropic nature, as reviewed by Maeda et al. (1977). These characteristics are expressed vividly in soils in Iceland. Vitric materials or Vitrisols (Icelandic system, Vitric Andosols of the deserts) do not show these properties as clearly as allophanic or metal-humus Andosols, but their physical behavior depends on their type and degree of weathering (see Warkentin and Madea 1980).

Andosols are light and fluffy soils. Low bulk density is one of the diagnostic criteria for Andosols with density of <0.9 g cm^{-3} required. The mineral colloidal fraction also

Fig. 5.8 Landslide on the Icelandic landscape. The soils easily reach the liquid limit (thixotropic) when disturbed when they are water saturated. Landslides are a very common phenomena in Iceland, especially where there are conditions for water charging the soils, from snowmelt (as in figure), or excessive rains when the soils are not frozen (spring and fall). Disturbance from heavy grazing animals can facilitate the triggering of landslides. Tephra layers and the bedrock below the soils commonly provide slip-planes

forms stable silt-sized aggregates that influence the physical properties of Andosols (Maeda et al. 1977) and make conventional mechanical particle size determinations useless for Andosols. Drying can cause irreversible decrease in water retention and increase in bulk density.

Andosols can retain large amounts of water, which is one of the main characteristics of such soils, hence the low bulk density. These soils commonly contain >60 % water (per dry weight of soil) at the wilting point for plants (15 bar, 1.5 MPa), which indicates their unusual water storage capacity. The term 'hydric' is used to describe Andosols when water retention is >100 % at 15 bar tension based on dry weight of the soil. While allophane, imogolite, and ferrihydrite contribute to this strong water retention the effect of organic matter (metal–humus complexes, allophane-humus and humus alone) is also important. Andosols have a large proportion of both large and intermediate pores, which allow for rapid water transport. Water infiltration and both saturated and unsaturated hydraulic conductivity are rapid compared to most other soils (see Warkentin and Maeda 1980; Basile et al. 2003). The aggregation of clay materials to silt-sized particles and the extremely high water retention leads to high frost susceptibility of Andosols (Arnalds 2004), which has a pronounced impact on soils of Iceland (see Chap. 10 on cryoturbation).

Andosols possess a special property called thixotropy. The soils can contain large amounts of water and yet appear relatively dry. When disturbed, the water is released. In other words, the soil can reach the liquid limit upon disturbance. This property is also expressed by very high liquid limits but a low range where the soil is plastic, resulting in very low plasticity index (often near 0). This property explains in part why Andosols are quite susceptible to slope failures when disturbed (Fig. 5.8).

5.8 Chemical Properties

Andosols can have a range of soil pH (measured in H_2O), in Iceland the range is typically 4.5–7.5. Metal-humus dominated soils tend to be acid (<5) with low base saturation. Soils dominated by allophane often have pH 5.5–6.5 (Nanzyo et al. 1993). If fresh basic parent materials are still present, which is typical in Iceland, pH is maintained by recharge of basic cations during weathering, which in Iceland sometimes leads to pH > 7 (Arnalds 2004). Soil reaction of Andosols rises rapidly when NaF is added to the soil solution, with F^- replacing OH^- from active surfaces. This is sometimes used to identify the presence of andic soil materials, both in the laboratory and in the field.

Andosols have pH-dependent charge as is common in tropical soils. Allophane, imogolite, ferrihydrite, and metal–humus complexes all have large reactive surface areas, but cation exchange capacity rises rapidly with increasing pH (see Wada 1985). Determination of CEC is therefore dependent on the pH used in any particular method (see Madeira et al. 2003). Common CEC values reported for Andosols range between 10 and 40 $cmol_c\ kg^{-1}$. Andosols also exhibit anion exchange properties that can be important for nutrient retention (e.g., Cl^-, NO_3^-, SO_4^{2-}).

Exchange characteristics make Andosols susceptible to heavy metal and radiocaesium (^{137}Cs) pollution (e.g., Adamo et al. 2003) by retaining the pollutants quite effectively, especially when the soils are not very acid (Nanzyo et al. 1993).

References

Regarding punctuation and Icelandic characters in citations: see note on punctuation in the Preface

Adamo P, Denaix L, Terribile F, Zampella M (2003) Characterization of heavy metals in contaminated volcanic soils of the Solofrana river valley (southern Italy). Geoderma 117:347–366
Arnalds O (2004) Volcanic soils of Iceland. Catena 56:3–20
Arnalds O (2013) The influence of volcanic tephra (ash) on ecosystems. Adv Agron 121:331–380
Arnalds O (2008) Andosols. In: Chesworth W (ed) Encyclopedia of soil science. Springer, Dordrecht, pp 39–46
Arnalds O, Kimble J (2001) Andisols of deserts in Iceland. Soil Sci Soc Am J 65:1778–1786

Arnalds O, Stahr K (2004) Volcanic soil resources: occurrence, development, and properties. Catena vol 56 (Special Issue). Elsevier, Amsterdam
Arnalds O, Oskarsson H (2009) A soil map of Iceland (Íslenskt jarðvegskort). Náttúrufræðingurinn 73:107–121 (English summary, table and figure legends)
Arnalds O, Orradottir B, Aradottir AL (2013) Carbon accumulation in Icelandic desert Andosols during early stages of restoration. Geoderma 193–194:172–179
Arnalds O, Bartoli F, Buurman P, Oskarsson H, Stoops G, Garcia-Rodeja E (eds) (2007) Soils of volcanic regions in Europe. Springer, Heidelberg, pp 469–491
Batjes NH (1996) Total carbon and nitrogen in the soils of the world. Eur J Soil Sci 47:151–163
Bartoli F, Buurman P, Delvaux B, Madeira M (2003) Volcanic soils: properties and processes as a function of soil genesis and land use. Geoderma, vol 117 (Special Issue). Elsevier, Amsterdam
Basile A, Mele G, Terribile F (2003) Soil hydraulic behaviour of a selected benchmark soil involved in the landslide of Sarno 1998. Geoderma 117:331–346
Bigham JM, Fitzpatrick RW, Schulze DG (2002) Iron oxides. In: Dixon JB, Schulze DG (eds) Soil mineralogy with environmental applications. Soil Science Society of America, Madison, pp 323–366
Dahlgren RA (1994) Quantification of allophane and imogolite. In: Amonette JE, Zelazny LW (eds) Quantitative methods in soil mineralogy. Soil Science Society of America Miscellaneous Publication, Madison
Dahlgren RA, Saigusa M, Ugolini FC (2004) The nature, properties and management of volcanic soils. Adv Agron 82:113–182
Duchaufour P (1977) Pedology (Paton TR). George Allen and Unwin, London
Eswaran H, van de Berg E, Reich P (1993) Organic carbon in soils of the world. Soil Sci Soc Am J 57:192–194
Fernandez CE, Yaalon DH (eds) (1985) Volcanic soils, weathering and landscape relationships of soils on tephra and basalt. Catena supplement 7
Gislason SR (2008) Weathering in Iceland. Jökull (Icel J Earth Sci) 58:387–408
Harsh J, Chorover J, Nizeyimana E (2002) Allophane and imogolite. In: Dixon JD, Schulze DG (eds) Soil mineralogy with environmental applications. Soil Science Society of America, Madison, pp 291–322
IUSS Working Group WRB (2006) World reference base for soil resources. World soil resources reports no 103. FAO, Rome
Jones A, Stolbovoy V, Tarnocai C, Broll G, Spaargaren O, Montanarella L (eds) (2010) Soil atlas of the northern circumpolar region. European Commission, Publications Office of the European Union, Luxembourg
Kimble JM, Ping C-L, Sumner ME, Wilding LP (2000) Andisols. In: Sumner ME (ed-in-chief) Handbook of soil science. CRC Press, New York, pp E209–124
Maeda T, Takenaka H, Warkentin BP (1977) Physical properties of allophane soils. Adv Agron 29:229–264
Madeira M, Auxtero E, Sousa E (2003) Cation and anion exchange properties of Andisols from the Azores, Portugal, as determined by the compulsive exchange and the ammonium acetate methods. Geoderma 117:225–241
McDaniel PA, Lowe DJ, Arnalds O, Ping C-L (2012) Andisols. In: Huang PM, Yuncong L, Sumner ME (eds) Handbook of soil sciences. Properties and processes, 2nd edn. CRC Press, Boca Raton, pp 33-29–33-48
Nanzyo M, Dahlgren R, Shoji S (1993) Chemical characteristics of volcanic ash soils. In: Shoji S, Nanzyo M, Dahlgren RA (eds) Volcanic ash soils. Genesis, properties and utilization. Developments in soil science, vol 21. Elsevier, Amsterdam, pp 145–187
Ndayiragije S, Delvaux B (2004) Selective sorption of potassium in a weathering sequence of volcanic ash soils from Guadeloupe, French West Indies. Catena 56:185–198
Parfitt RL, Clayden B (1991) Andisols—the development of a new order in Soil Taxonomy. Geoderma 49:181–198
Parfitt RL, Kimble J (1989) Conditions for formation of allophane in soils. Soil Sci Soc Am J 53:971–977
Shoji S, Takahashi T (2002) Environmental and agricultural significance of volcanic ash soils. Global Environ Res 6:113–135
Shoji S, Nanzyo M, Dahlgren RA (1993) Volcanic ash soils. Genesis, properties and utilization. Developments in soil science vol 21. Elsevier, Amsterdam
Shoji S, Nanzyo M, Dahlgren RA, Quantin P (1996) Evaluation and proposed revisions of criteria for Andosols in the World Reference Base for Soil Resources. Soil Sci 161:604–615
Shoji S, Toyoaki I, Saigusa M, Yamada I (1985) Properties of nonallophanic Andosols from Japan. Soil Sci 140:264–277
Soil Survey Staff (1999) Soil Taxonomy. A basic system of soil classification for making and interpreting soil surveys. In: Agricultural handbook no 436. USDA, NRCS, Washington DC
Soil Survey Staff (2003) Keys to Soil Taxonomy, 9th edn. USDA, NRCS, Washington DC
Smith G (1986) The Guy Smith interviews: rationale for concepts in Soil Taxonomy. In: SMSS technical monograph no 11. Cornell University, Boston
Schwertmann U (1985) The effect of pedogenic environments on iron oxide minerals. In: Advances in soil science vol 1. Springer, New York, pp 172–200
Stefansson A, Gislason SR (2001) Chemical weathering of basalts, southwest Iceland: effect of rock crystallinity and secondary minerals on chemical fluxes to the ocean. Am J Sci 301:513–556
Tan KH (1984) Andosols. In: Benchmark papers in soil science series. Van Nostrand Reinhold Company, New York
Torn MS, Trumbore SE, Chadwick OA, Vitousek PM, Hendricks DM (1997) Mineral control of soil organic carbon storage and turnover. Nature 389:170–173
Wada K (1985) The distinctive properties of Andosols. Adv Soil Sci 2:173–229
Wada K (1989) Allophane and imogolite. In: Dixon JB, Weed SB (eds) Minerals in soil environments, 2nd edn. Soil Science Society of America, Madison, pp 1051–1087
Wada K, Arnalds O, Kakuto Y, Wilding LP, Hallmark CT (1992) Clay minerals of four soils formed in eolian and tephra materials in Iceland. Geoderma 52:351–365
Warkentin BP, Maeda T (1980) Physical and mechanical characteristics of Andisols. In: Theng VKG (ed) Soils with variable charge. New Zealand Soil Bureau, Lower Hutt, pp 97–107
White GN, Dixon JB (2002) Kaolin–serpentine minerals. In: Dixon JD, Schulze DG (eds) Soil mineralogy with environmental applications. Soil Science Society of America, Madison, pp 389–414
Wolff-Boenisch D, Gislason SR, Oelkers EH, Putnis CV (2004) The dissolution rates of natural glasses as a function of their composition at pH 4 and 10.6, and temperatures from 25 to 74 °C. Geochim Cosmochim Acta 68:483–4858
Zehetner F (2010) Does organic carbon sequestration in volcanic soils offset volcanic CO_2 emissions? Quatern Sci Rev 29:1313–1316

6 Classification and the Main Soil Types

6.1 Introduction and Historical Notes

The Andosol soil group of the WRB and the Andisol order of the Soil Taxonomy were developed before information about Icelandic Andosols was reported extensively in the literature. These systems are poorly suited for Icelandic Andosols, especially Soil Taxonomy with its climatic bias, resulting in contrasting soils being classified as the same soil on sub-order and even great group level (Vitricryands). The WRB is better suited for soils of Iceland, but still fails separating the desert soils from the fertile Brown Andosols. The discussion in this book is based on an Icelandic classification scheme. The major difference between the WRB and the Icelandic system is that it separates the desert soils as a special soil group, Vitrisols (Fig. 6.1). This is absolutely necessary considering the dramatic difference between the desert soils (Vitrisols) and the soils under vegetation (Andosols). In addition, deserts cover >40 % of Iceland, which justifies this separation still further. The current level of the system was published by Arnalds and Oskarsson (2009), but is also explained in English in a paper in the journal Jökull (Arnalds 2008). A soil map of Iceland is published at the end of this chapter.

The oldest soil map of Iceland that the author has come across is from the early twentieth century published by a German named Gruner (1912). It is simple and seems largely based on earlier geodesic mapping by the Danish military (Iceland was under Danish rule). The Danish land survey in 1:50,000 differentiated between vegetated land and barren and also took notice of wetlands, a separation that also set apart the main soil categories. The first major attempt to systematically map the soils of Iceland was undertaken by the US soil scientist I.J. Nygard (Nygard and Johannesson 1959) (Fig. 6.2), but the soil classification for the map is described in a monograph by Johannesson (1960) titled "The Soils of Iceland." The soil map was included with his monograph. Johannesson's book was also published in Icelandic. The map is in the scale 1:750,000 and it gives a good overview of the soils of Iceland. A closer look at the map suggests that the map polygons are, at least in part, based on Danish land survey maps. The Icelandic version of the book was re-issued in 1988, with an extended appendix reviewing soil research up to the publication date (Johannesson 1988).

Johannesson made a distinction between soils of deserts and soils under vegetation, a separation that is still being used under the current system (see later in the chapter). Vegetated land was divided into *Peat soils* (wetlands), *silty soils* (heathland dominating) and *gravelly and rocky soils*. The deserts (soils with little or no vegetation) were divided into *sands*, and *gravelly and stony materials*. The total number of soil mapping units was 20 and 3 additional units represented surfaces with "neither soil nor vegetation." Soil thickness, coarse fragments (stoniness), and the nature of the underlying materials were factors used to differentiate between the soil mapping units. The mapping is undertaken in a coarse scale, so each map unit describes predominant soils, and associated soils are also noted with a broad estimate of aerial proportion for each of the categories. An example is mapping unit 1, *peat on gravel and sand* (40–60 %), but associated soils are *silt loams* (20–40 %) and other kinds of soils (15–25 %). Soil associations dominated by vegetated land are about 29,000 km^2, but subtracting the estimated extent of barren soils within these associations gives a sum for soils under vegetation cover at about 24,000 km^2. This is a considerably lower number than indicated by the AUI Farmland Database (Nytjaland, Chap. 4), but closer to results from the Danish geodesic mapping and often cited 25 % vegetation cover during the latter part of the last century. It is, however, difficult to calculate the total aerial extent of each soil type because each unit is an association of soil types. The peat soils correspond to Histsols, Histic Andosols, and Gleyic Andosols in the current system, but it should be noted that a large part of the peat soils under the Nygard and Johannesson's mapping do not qualify as Histosols (true peat soils) due to low organic content of many of the Icelandic wetlands. The silt loams are predominantly Brown Andosols and soils without vegetation

O. Arnalds, *The Soils of Iceland*,
World Soils Book Series, DOI 10.1007/978-94-017-9621-7_6
© Springer Science+Business Media Dordrecht 2015

Fig. 6.1 There is a dramatic difference in soil characteristics and ecosystem functions between soils under vegetation (Andosols) and desert soils (Vitrisols). These differences are not adequately accounted for under the Soil Taxonomy or WRB, but separation of these soils at the highest level is an important attribute of the Icelandic classification scheme. The soil on the *left* is a Brown Andosol, a fertile agricultural soil. It shows marked signs of cryoturbation with *light colored* rhyolitic Mt. Hekla tephralayers. The soil on the *right* is a Sandy Vitrisol, formed in recent glacio-fluvial deposits, lacking both clays and organic matter

would be Vitrisols under the current system. Andosols were not yet recognized by the science community at the time when this work was undertaken, which limits its usefulness today. And, naturally, better geographic data for Iceland is now available. Yet, the soil map and the monograph "Soils of Iceland" are monumental achievements worth commemorating.

Gudmundsson (1994) translated and adapted the FAO soil classification from 1988 (FAO-UNESCO 1988, predecessor to the current WRB) for the soils of Iceland. The main soil types were Andosols, Histosols, Arenosols, Fluvisols, Gleysols, Regosols, and Leptosols. However, much of the Histosols, Arenosols, Fluvisols, Regosols, and Gleysols under this Icelandic version of the 1988 FAO legend are now classified as Andisols (Soil Taxonomy) or Andosols (current WRB), as they meet the criteria for andic or vitric properties according to the WRB and andic soil properties according to Soil Taxonomy (Arnalds et al. 1995; Arnalds and Kimble 2001; Arnalds 2004; Arnalds and Oskarsson 2009). Gudmundsson's work was, however, a remarkable step forward, recognizing Andosols in Iceland within his system, and some of the principles he developed are used in the current system presented below.

6.2 Main Classes

The system used here takes notice of the current WRB soil groups (IUSS Working Group WRB 2006). It separates between Histosols, Andosols, Vitrisols, and other soils at the highest level. The use of the term Vitrisols is not unique to Iceland; it is also used under the French classification system (INRA 1998) and Pumice soils are a special unit under the New Zealand classification (Hewitt 1993, 1998). This distinction is important for the soils of Iceland and is worth considering for use elsewhere. The reason is that tephra (volcanic ash) as parent material is quite unique and differs from other sandy parent materials in that it has considerable surface area, leading to positive soil properties such as water retention and CEC. Furthermore, basaltic tephra weathers rapidly, and even though the materials are relatively unweathered, basic cations are released at a considerably fast rate and retained by exchange sites, making the Vitrisols a fertile medium for plant growth, if other environmental factors allow for it. This is less evident for silicious tephra compared to basaltic volcanic materials. Much of the basaltic tephra meets the criteria for vitric materials (andic soil

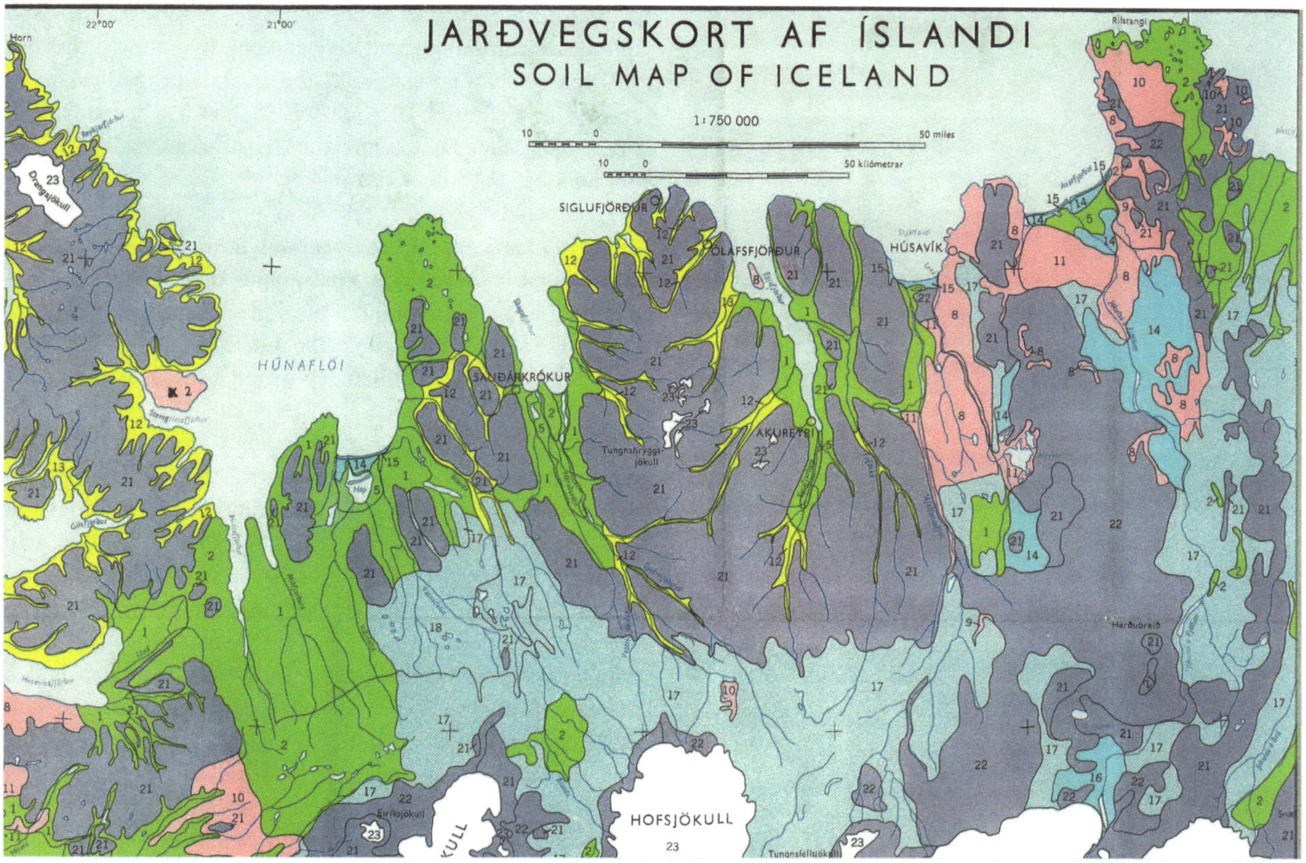

Fig. 6.2 Part of the Nygard/Johannesson soil map from 1959. Soils with vegetation cover are *green* (wetlands dominate), *pink* (drylands) and *yellow* (gravelly soils). Deserts are *bluish*, with "mountainous land with neither soil nor vegetation" shown as *dark blue-grayish*

properties under Soil Taxonomy) when it is freshly deposited. This indicates the level of physical and chemical properties the basaltic tephra has as a fresh parent material, which is in no way equivalent to sandy materials made of quartz or carboniferous materials. Vitric Leptosols/Entisols or Vitric Inceptisols/Regosols (Soil Taxonomy and WRBH) are simply not comparable to non-weathered quartz or rocky materials with the same classification. This alone justifies the use of the term Vitrisol, not just in Iceland, but on an international scale. Furthermore, deserts with surfaces dominated by tephra materials are dominating large proportions of Icelandic landscapes, which need to be differentiated from other soil surfaces at the highest level.

The separation into major soil classes is shown in Fig. 6.3, the diagnostic criteria in Table 6.1.

The diagnostic criteria are made as simple as possible. It considers the weighted average of the top 30 cm, but there is no minimum depth limit. It should be noted here, as was in the previous chapter, that Andosols can have up to 25 % C according to the WRB, compared to 12–18 % limit between Histosol and other soil groups/orders. The WRB limit was recently lifted from 20 to 25 % (same as Soil Taxonomy). The Icelandic system still uses the 20 % C limit and it seems sensible to do so to allow more space for the organic peat soils, also considering that these soils generally have more carbon in horizons under the 30 cm control section. Allophane is determined based on Si_{ox} content ($Si_{ox} \times 6$).

Below is a discussion of the soil classes with figures illustrating examples of each of the soil types. The soil map is shown at the end of the chapter, but examples of soil descriptions and analytical data are presented in the subsequent chapters.

6.3 Andosols

Icelandic Andosols are soils with andic soil properties that occur under vegetation (hence have organic carbon above the minimum of 1.5 % C in the surface horizon). They are separated from Vitrisols based on having more carbon and allophane than the Vitrisols (Table 6.1). Allophane is determined based on Si_{ox} content. The upper carbon level is

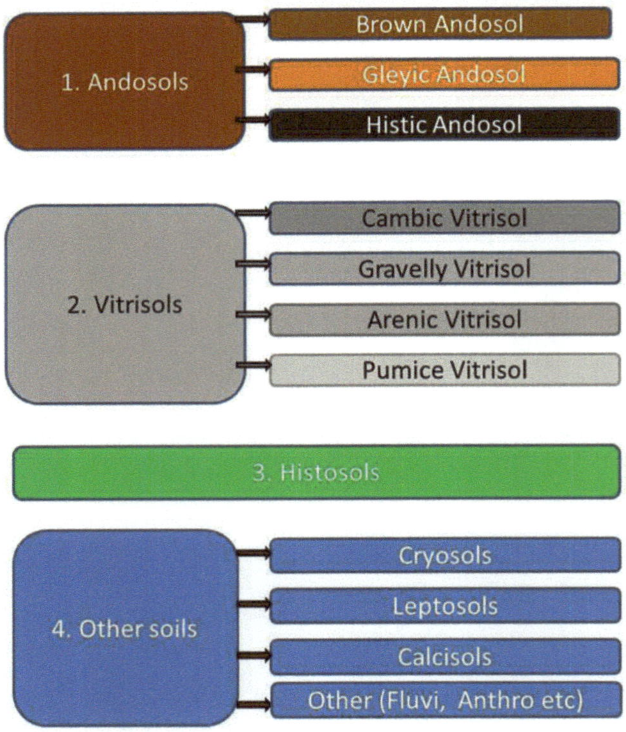

Fig. 6.3 Schematic visualization of the Icelandic classification. Four main categories are split into 11 main soil types plus "other soils"

20 % C, but Histosols are classified by >20 % C in surface horizons. The Andosols can be shallow over vitric materials, when it has an A horizon meeting the allophane and carbon criteria.

6.3.1 The Pedogenic Parameters Underlying the Separation of Andosols

The separation between the Andosol classes (or soil types) is based on dominant influences of (1) amount of steady aeolian input and (2) drainage category. These factors determine if the soils are wetland or dryland soils, but the aeolian input influences carbon content, clay content, hydraulic properties, soil reaction, grain-size, and the overall properties of the soils. The separation of Andosols is shown schematically in Fig. 6.4, together with the Histosols and Vitrisols boundaries.

The steady aeolian deposition is one of the most distinctive characters of the soil environment. The aeolian sedimentation rates are commonly 0.01–1 mm year^{-1}, resulting in a steady burial of the soil surface (see Chap. 9 on soil genesis). Periodic volcanic additions and momentary augmentation of the aeolian deposition rates, as witnessed by recent volcanic eruptions in Eyjafjallajökull (2010) and Grímsvötn (2011) enhance this effect (see Arnalds 2010; Arnalds et al. 2013). A discussion of the aeolian environment, a major factor affecting Icelandic geomorphology and soil environments is provided in Chap. 11, with a map showing the geographic distribution of the deposition.

As the aeolian input decreases with more distance from the main aeolian sources (Fig. 6.4, *y*-axis), the soil thickening rates are reduced. Finer sediments are deposited at distance from the aeolian sources compared to deposition close to the aeolian sources. The slow deposition rate and finer materials means enhanced soil development per depth increment. That translates into more accumulation of organic materials and increased formation of clays (allophane and ferrihydrite) for each depth increment of the soil pedon. Wetland positions further enhance carbon accumulation, resulting in Histosols and Histic Andosols in wetlands far from aeolian sources, such as in West and North Iceland (organic arrow at the center of Fig. 6.4 pointing to the lower left corner). However, with increased aeolian deposition, the mineral materials become a more substantial proportion of the soil; the carbon level drops below 12 % and the soil becomes Gleyic Andosol. Many Gleyic Andosols within the most active aeolian areas have less than 3 % carbon in surface horizons. With more rapid aeolian influx, less time is given for soil formation before the soil is buried under new

Table 6.1 The principal soil classes and corresponding terms of the Soil Taxonomy and the WRB

Soil class	Symbol	Identification	S.T.	WRB (2006)
Histosol	H	>20 % C	Histosol	Histosol
Histic Andosol	HA	12–20 % C	Aquand	Histic and Vitric Andosol
Gleyic Andosol	GA	<12 % C; gleying/mottles	Aquand	Gleyic, Histic and Vitric Andosol
Brown Andosol	BA	<12 % C, dry; >6 % allophane	Cryand	Vtiric, Silandic Andosol and more
Cambic Vitrisol	MV/GV	<1.5 % C; <6 % allophane	Cryand	Vitric Andosol/Regosol/Leptosol
Arenic Vitrisol	SV	Sand, <1.5 % C	Cryand	Vitric Andosol/Arenosol/Leptosol
Pumice Vitrisol	PV	Pumice >2 mm	Cryand/Entisol	Regosol/Vitric Andosol
Leptosol	L	Rock/scree	Entisol	Leptosol
Cryosol	C	Permafrost	Gelisol	Cryosol

Identification criteria also shown. Table slightly modified from Arnalds and Oskarsson (2009)

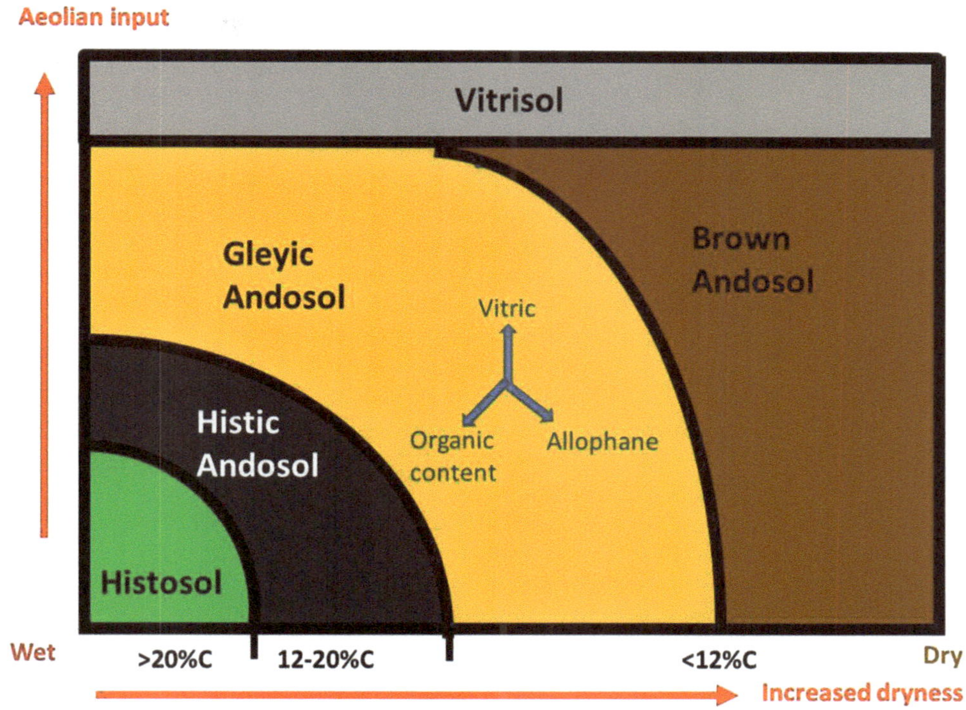

Fig. 6.4 Separation of Icelandic Andosols, as influenced by drainage and aeolian sedimentation rates. Histosols and Vitrisols also shown. 20 % C separates Histosols, but Vitrisols are low in allophane and organic materials

aeolian materials; younger soils on top of buried soils. However, as the organic factor takes over (little aeolian deposition in wetland positions), clay formation slows down, concurrent with lower pH values. Allophane does not form at pH < 4.9 (see Chap. 8). Clay formation is also slower for each depth increment when the aeolian deposition is quite rapid. Consequently, ideal conditions for clay formation are found in dryland positions where aeolian deposition is not rapid (allophane axis in Fig. 6.4 pointing to the lower right corner). The vitric axis is pointed with the aeolian deposition axis (y-axis, toward the top of Fig. 6.4). These three axes are concurrent with the three major types of Andosols in general, as was discussed in the previous chapter, namely the alu-andic (organo-mineral, organic), sil-andic (allopanic), and vitric soils.

The trends for the various soil properties influenced by the amount of aeolian deposition are presented in Fig. 6.5. The x-axis shows increased aeolian deposition, but the y-axis shows % C and % clay (on left), and pH (on right), while the lines depicted are indicative of other soil properties such as

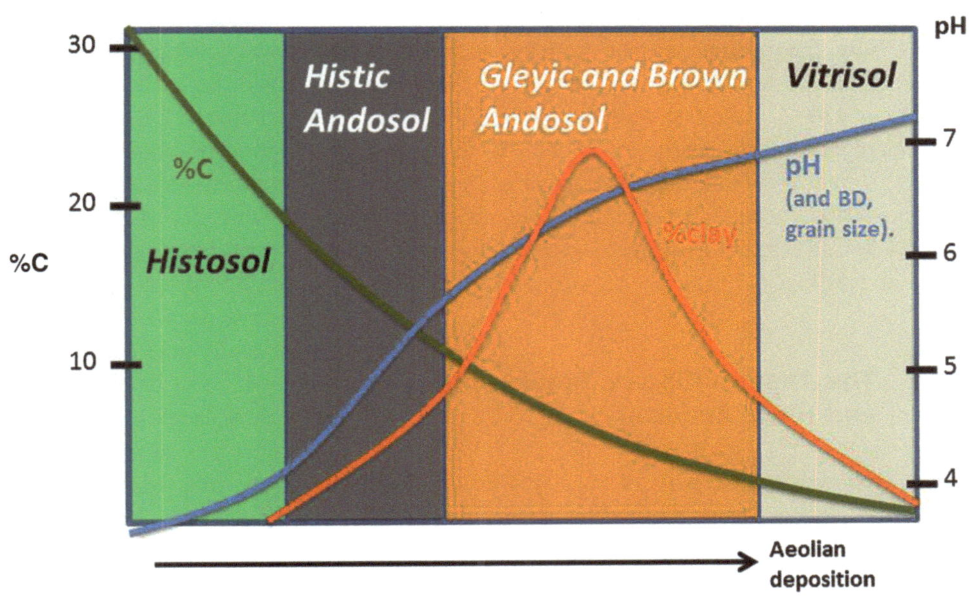

Fig. 6.5 The influence of aeolian deposition on clay content, pH and organic carbon. Histic Andosols and Histosols form in wetland landscape positions

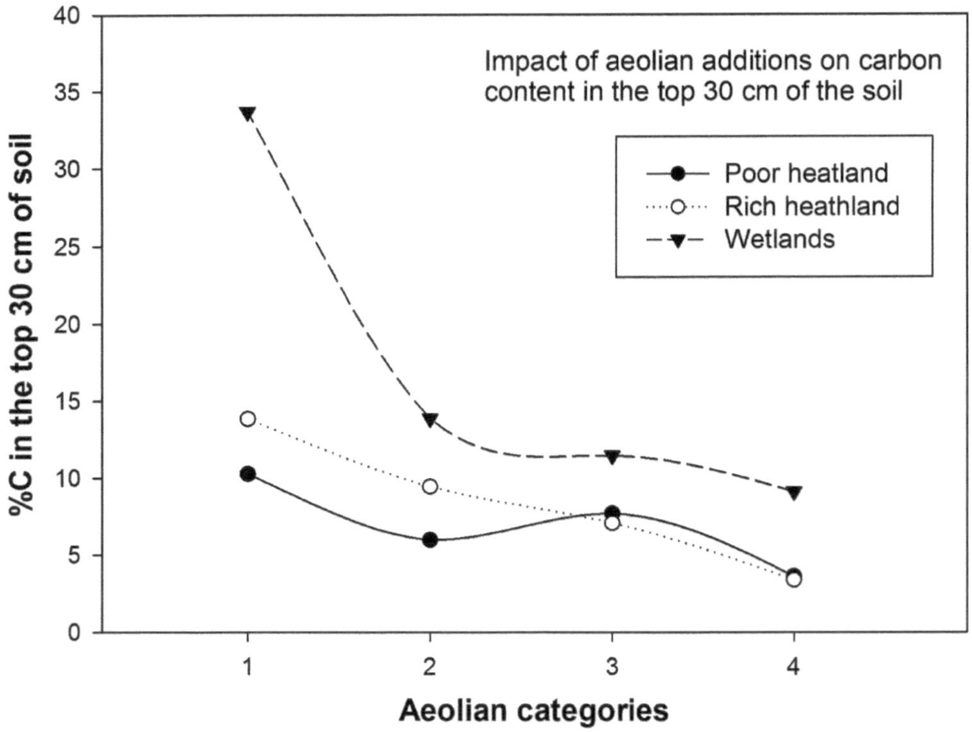

Fig. 6.6 The relationship between the level of aeolian additions (*x*-axis) and carbon in the top 30 cm of soils. *1* least aeolian deposition; *4* rapid aeolian deposition. Based on Agricultural University of Iceland IGLUD database (random samples, total of 63 for wetlands, 116 for rich heathland, and 257 samples for poor heathland (see Chap. 4 on vegetation). © AUI IGLUD/JG/OA

bulk density and grain size (blue line). Maximum clay accumulation for each depth increment can be expected where aeolian deposition is neither very rapid nor slow, in both dryland and wetland positions. Soil reaction (pH) drops with reduction in aeolian inputs, and the carbon levels generally rise. Inorganic soils dominated by the aeolian influences are Vitrisols (to the right on the graph).

The effect of aeolian deposition on carbon content in wetlands and heathlands is shown in Fig. 6.6. It shows clearly how the average carbon content decreases from >30 % C in the top 30 cm in wetlands far from aeolian sources to <10 % closest to aeolian sources. The trend for rich heathland is from nearly 15 % C far from the aeolian source to <5 % C within the most active aeolian area. The carbon in poor heathland decreases from about 10 % to below 4 % C. These trends correspond to the line for C in the figure above (Fig. 6.5).

6.3.2 The Andosol Classes: Brown, Gleyic, and Histic Andosols

Brown Andosols are the soils of vegetated drylands. These are often typical of Andosols, in many ways similar to those found in Massif Central in France, Hokkaido Japan, or in Alaska, as represented by the amount of allophane and organic matter, and physical properties such as water retention. Incidentally, the term "Brown Andosol" is also used for the soil classification in Hokkaido (Hokkaido Soil Classification Committee 1979). However, the active aeolian deposition makes these soils coarser close to aeolian sources. The soils contain many volcanic tephra (ash) layers, because of the frequent volcanic activity (see Chap. 3). The tephra layers are more distinct and coarse close to the most active volcanoes, such as Hekla and Katla, but some tephra layers cover a large proportion of Iceland, especially the silicic layers from Hekla. The two pedons in Fig. 6.7 are typical examples of Brown Andosols. Both are about 1 m deep, resting on about 10,000 year old glacial till. Both receive intermediate amounts of aeolian additions. The pedons have organic rich A horizons and carbon content remains high, with quite erratic distribution throughout the profiles (see descriptions at the end of the next chapter). Plant roots extend deep into the profiles, but the soils are light (BD generally <0.8 g cm^{-3}) and "fluffy". The soils are easy to dig in with a spade. The color of the soil to the left is redder and it shows some signs of gleying in the lower horizons, but it is sometimes saturated during winter and spring when frozen, but is otherwise freely drained. Both soils show signs of more rapid aeolian sedimentation rates over the past 1,100 years after the Settlement of Iceland. Hummocks are common on the surface.

Fig. 6.7 Two examples of Brown Andosols. The one on the *left* is from E Iceland. The *lower half* has some clay loam horizons but the *upper half*, deposited after the Settlement (*darker part*) is typical silt loam. Traces of dark and light colored tephra layers can be seen. The pedon on the *right* is from South Iceland with numerous tephra layers, mainly from Mt. Hekla

The pedon to the right (Fig. 6.7a) from South Iceland exhibits thick tephra layers, especially from Mt. Hekla. This pedon is about 50 km away from the Hekla volcano. It shows many features caused by cryoturbation. The soil on the left, from East Iceland, does not have as clear tephra layers although some traces of both dark and light colored tephra can be seen. It is also a typical Brown Andosol with considerable carbon, allophane, and ferrihydrite content from top to bottom, with erratic distribution.

Gleyic Andosols are the soils of wetlands with <12 % C in surface horizons. They show strong andic soil properties, as the Brown Andosols, expressed by numbers for $(Al + ½ Fe)_{ox}$ far exceeding the 2 % limit. Allophane and ferrihydrite contents are similar to those for Brown Andosols, often 10–20 % allophane and 3–8 % ferrihydrite. Tephra layers are common, as many of these soils occur in proximity of active volcanoes, and also receive considerable aeolian deposition.

Two Gleyic Andosol pedons are presented in Fig. 6.8 but descriptions and some other characteristics are presented at the end of the next chapter. The one on the left is obtained near the town Hella in South Iceland, and it is one of the so-called COST-622 profiles, a part of an EU sponsored collaborative research and cooperation on volcanic soils of Europe. The COST-622 profiles were subjected to a range of scientific analysis in many European laboratories (see Arnalds et al. 2007). This profile shows many of the characteristic features of Gleyic Andosols. The profile is relatively thick (>2 m), with a pronounced difference in properties between surface horizons and subsurface horizons, with the lower part being buried Histic Andosols and Histosols. There is a clear sign of the Settlement of Iceland in the strata, signified by a distinct color difference with lighter colored soils formed after the Settlement. The time of the Settlement is also easy to identify by the two-color Settlement layer dated from time of Settlement (about 874 AD), which is shown in Fig. 6.9 (close-up from the profile). The lighter color above the Settlement is caused by increased aeolian sedimentation rates (fourfold to tenfold the pre-Settlement rates, see Chap. 12), with a proportion of rhyolitic light colored tephra grains. This increased aeolian sedimentation rates results in lower carbon contents for each depth increment, resulting in the Gleyic Andosol classification, but Histic Andosols are also common in the southern lowlands. It is clear that Histosols were much more common in this area at the time of Settlement than they are now. However, pollen research has shown a warm period with extensive woodlands 8,500–6,000 years ago (Hallsdottir and Caseldine 2005), but tree trunks from birch are often found deep in the soils of the current wetlands. Wetlands expanded gradually after this main birch-period (see also Norddahl et al. 2008), but birch expanded again in some places, e.g., in the south, to peak before the Settlement of Iceland (Hallsdottir and Caseldine 2005). The pedon to the right is obtained from a wetland in Hornafjörður, Southeast Iceland. It is much shallower, most likely a new soil that has formed on previously disturbed site (disturbed by a flooding glacial river). The site receives large amount of aeolian sediments, resulting in the platy structure, each band often representing winderosion events or series of events nearby. The carbon content is low but still above the required 1.5 % C.

(a) (b)

Fig. 6.8 Gleyic Andosols. **a** Deep wetlands soil to the *left* (South Iceland) and shallow wetland soil to the *right* (Southeast Iceland). The Gleyic Andosol on the *left* overlies a buried Histosol. A clear color difference, manifested by the Settlement tephra layer occurs at the *middle* of the profile, resulting from increased aeolian deposition after the Settlement. Clear signs of cryoturbation in the *upper part* of the profile. **b** The profile on the *right* receives rapid aeolian deposition, depicted by the platy structure

Cryoturbation is evident in surface horizons in Fig. 6.8a, imprinted by the waving dark tephra layer from the Middle-Ages. It is worth noticing that older horizons are not cryoturbated, reflecting both change in surface vegetation cover and climatic change, with the climate becoming cooler during the Middle-Ages; (see Chap. 10 on frost and cryoturbation).

Features showing gleying and mottles are common in Gleyic Andosols as can be expected. Some examples are shown in Fig. 6.10.

Histic Andosols are Andosols with 12–20 % C in the top 30 cm of soils (weighted average). The majority of these soils are wetlands (saturated and damp wetlands according to the AUI Farmland Database, see Chap. 4), but some are dryland soils including rich heathlands, birch forests and grasslands far from aeolian sources. Examples of Histic Andosol from West and North Iceland are shown in Fig. 6.11.

Fig. 6.9 A close-up of the Settlement strata, seen in the Hella COST-622 profile. Scale on the *right* is in centimeters. The Settlement layer is distinctive. The soils above the Settlement continue to be organic in nature, but the organic content gradually decreases toward the *top* of the profile

Fig. 6.10 Oxidation features in Gleyic Andosols. To the *left* is a red oxidized zone that follows a coarse tephra layer (better aeration) and some thinner layers showing the same features. To the *right* are oxidized (*red*) areas around roots of wetland plants. Both *photos* were taken near the Hálslón Reservoir, in the eastern highlands

Fig. 6.11 Histic Andosol from West Iceland (*left*) and North Iceland (*right*). The West Iceland example is acidic (pH 4–4.5) and relatively thin (<1 m), with layers of organic materials. The North Iceland example is a deep soil, with distinct Mt. Hekla rhyolitic tephra layers. It has higher pH (>5) and a growing organic content with depth. The top has <20 % C because of increased aeolian activity after the Settlement and improved drainage in the surface

(a)

(b)

The Histic Andosol from West Iceland occurs in relatively dry landscape position far from aeolian sources, but to the right is a Histic Andosol in wetland position receiving intermediate levels of aeolian inputs. These soils are a peculiar mixture of histic and andic soils, with the andic soil properties dominating. Distinctive organic build-up phases can clearly been seen in the West Iceland profile (left), but oxidation features around a coarse tephra layer are prominent near the base (130 cm depth).

The Histic Andosol on the right is located at the AUI experimental farm at Möðruvellir in North Iceland. It is nearly 4 m deep. It shows clearly the H3 and H4 tephra layers from Hekla. The H1 (1104 AD) layer is cryoturbated and less distinctive about 20 cm above H3. Under the Histic Andosols are buried Histosols, hence the dark to black color of the lowest horizons. Some of the deep horizons have preserved birch and/or willow stems up to 6 cm in diameter. The soil is lighter in color toward the top as is common, due to increased aeolian deposition which partly contains light colored tephra grains (from erosion of soils upwind) and also because of lower organic matter content. This soil showed strong hydrophobic character on drying.

6.4 Histosols

Histosols are soils with more than 20 % C in the uppermost horizons (30 cm weighted average) according to the Icelandic classification scheme (Arnalds and Oskarsson 2009). Organic Histosols are only found where aeolian deposition rates are low. They are mainly found in the westernmost and northernmost parts of Iceland. Prominent areas with Histosols are found on 'Mýrar' which are the wetland areas close to the coast in West Iceland, on the Snæfellsnes peninsula and the Westfjord penisula and the Skagi and Tröllaskagi peninsulas in the North. An example of Histosols from Dýrafjörður in the Westfjords is shown in Fig. 6.12.

Total extent of Histosols is rather limited. Many early discussions of Icelandic wetlands soils did not separate aquic soils with low organic content (Gleyic Andosols and Histic Andosols) from the peat soils with high organic content (Histosols), which can make review of the literature confusing. It should be noted here that many of the Histosols have subsurface horizons with higher organic content than the surface horizons, as was explained above for Gleyic and Histic Andosols.

There are several publications that provide overview of Icelandic peat soils and wetlands. The Ph.D. thesis of Gudmundsson (1978) gave a detailed account of some Histosols in West and Northwest Iceland, however unpublished. An accumulation of papers on Icelandic wetlands were presented in a book in Icelandic (Olafsson 1998). Johannesson's monograph (1960) on the soils of Iceland has several sections dealing with the soil group "peat soils" which entails Gleyic and Histic Andosols together with Histosols. In addition, useful information can be drawn from the various papers on the Icelandic pedons of the European COST-622 study (see Arnalds et al. 2007). Bjarnason (1952, 1966) conducted a survey on the Icelandic peat with its possible use as fuel as the main theme.

The geological separation of wetlands (and then sometimes collectively termed peat even though not all are peat) are (i) "flói" which are broad level wetlands; and (ii) hallamýri, or wetlands on slopes, where the ground-water seeps downwards and the water level reaches the surface (Einarsson 1968). The "flói" is considered to be chiefly rainfed while the "hallamýri" is charged by rainwater, runoff and

Fig. 6.12 Histosol from Dýrafjörður in the Westfjords. A newly dug ditch in a gently sloping wetland. All horizons are organic, the soil has low pH of about 4 and limited allophane content

groundwater (Gudmundsson 1978). However, a mixture of both is very common.

The carbon content reaches 40 % in some horizons, but is generally much lower (20–30 %). The main character of the organic matter in the soils is that it is poorly decomposed and the soils would classify as Fibrists (Borofibrists and Cryofibrists) under Soil Taxonomy. They show considerable shrinkage when they are completely dried in the laboratory (often less than half the original volume when dry), but limited or very slow shrinkage when drained, which is in part attributed to the volcanic ash materials in the matrix (Bartoli and Burtin 2007).

The Histosols do not contain appreciable amounts of allophane clays, as pH is generally low, but allophane formation is inhibited at pH below 4.9. The pH of the pedon shown in Fig. 6.12 is about 4. Still, the soils have some andic properties with a considerable amount of aluminum–humus complexes.

6.5 Vitrisols—The Andic Soils of the Deserts

The two most commonly used international classification systems, the WRB and the US Soil Taxonomy fall short in separating Icelandic desert soils from other soils, as was mentioned previously. Thus, Brown Andosols formed under vegetation classify the same as the very contrasting soils of the deserts at the higher level according to Soil Taxonomy (Vitricryands) and WRB (Vitric Andosols). There is a need for separating these soils, which have very extensive coverage (>40 % of Iceland) from other soils, as they have very different soil properties (to say the least), genesis and ecosystem services. They lack the high organic matter content and water holding capacity which is typical of Andosols and have very low allophane content. Their separation from other Andosols is fundamental to the soil map of Iceland. Yet, they meet the criterion for andic/vitric materials, which should be kept in mind in relation to their classification according to Soil Taxonomy and WRB.

The very nature of Icelandic Vitrisols makes one think how soils are defined. Are the Vitrisols soils or just sediments? Many Icelanders find it difficult to accept that deserts have a soil cover, which is understandable in light of the large difference between the organic Andosols and the poor Vitrisols. The author of this book has also been in correspondence with foreign scientists who have had the same difficulties and even made strong arguments that the less developed Andosols are not soils, but only sediments. Well, of course all these surfaces do have soils, but some of them are very poorly developed, just as many of the Entisols under the US Soil Taxonomy.

The Icelandic Vitrisols are divided into four subclasses: Cambic, Gravelly, Sandy (Arenic), and Pumice Vitrisols. These subclasses reflect the different geologic environments where Vitrisols occur. A comprehensive publication of the Vitrisols was published by Arnalds and Kimble (2001), who gave detailed analytical data for eight desert soil pedons. Other papers dealing with Vitrisols include those of Johannesson (1960), Arnalds (1988, 1990), Gudmundsson (1991), and Arnalds et al. (1995). The following discussion on Vitrisols is largely based on the paper by Arnalds and Oskarsson (2009) on soil classification.

The *Cambic Vitrisols* are typical of glacial till environments with rather low sandy aeolian inputs (Fig. 6.13). There is enough pedogenesis to cause color changes in subsurface horizons to create a cambic horizon (Bw). The clay in the Bw horizon can be formed in situ, it can be a remnant of a former Andosol cover that has been removed by soil erosion (common), or it can also have been deposited by aeolian processes caused by wind erosion of Andosols upwind. It is common to have 0.2–1 % organic carbon in the Bw horizon, but the surface is often more sandy with less organic carbon.

Fig. 6.13 Cambic Vitrisol. A shallow soil with a cambic B-horizon with about 1.2 % C. Frost-heave maintains the gravelly surface. The soil is silty and subject to intense cryoturbation

The surface is commonly gravelly, which is maintained by frost heaving. As a result of some organic matter and clay, the soils have considerable water retention and CEC, yet they are nutrient limited and the availability of the N is probably limited as the organic matter is strongly complexed as andic soil materials, represented by $(Al + \frac{1}{2}Fe)_{ox}$ ranging between 1 and 2 %. The fine materials enhance cryoturbation and sorting by freeze–thaw cycles, resulting in patterned ground on the surface (see Chap. 10).

The *Gravelly Vitrisols* have similar features as the Cambic Vitrisols, but are more gravelly throughout and lack a distinct cambic horizon. They are typical of raised beaches, such as in extensive areas in West Iceland that have lost the soil cover due to land degradation. They occur also where powerful glacial floods have deposited gravelly sediments, such as at the Markarfljótsaurar plains west and south of Eyjafjallajökull in South Iceland. They are also typical of fluvial plains along many rivers. These materials tend to have higher bulk density than the Sandy and Cambic

Vitrisols, and are less fertile as they have smaller component of fine materials <2 mm. However, they have sometimes accumulated fine silty materials from aeolian deposition.

The *Sandy Vitrisols* (also termed *Arenic Vitrisols*) are quite extensive in Iceland (>15,000 km^2) and are characterized by poorly weathered basaltic glass and rock fragments (Fig. 6.14). The thickness of these deposits varies from several meters to very thin layers over various geologic surfaces such as lava and till. In many cases there is a gradient from the Cambic Vitrisols to Sandy Vitrisols reflecting the various amount of aeolian deposition. As sand advances over lava surfaces (see Chap. 11 on aeolian environments), depressions become full of sand while rocks outcrop at topographic heights, making a complex mosaic of Sandy Vitrisols and Leptosols. However, the Icelandic system presented here does not put depth requirement on the Vitrisols, making most of these surfaces as Sandy Vitrisols. In spite of being young deposits or surfaces, these basaltic materials have acquired andic soil properties, with $(Al + \frac{1}{2}Fe)_{ox}$ commonly >1 %. These soils are characterized by very low organic contents (often <0.1 % C) and extremely unstable surfaces. Patterned ground is not as common on sandy surfaces as they are on Cambic Vitrisol surfaces.

Fig. 6.15 Pumice Vitrisol in pumice sediments close to Mt. Hekla. Note the finer soil below the new tephra. A sequence of tephra and buried soils is often found close to the volcanoes. *Photo* © Elin Fjola Thorarinsdottir (ISCS)

Fig. 6.14 Sandy Vitrisol beneath a gravelly surface. The location is near an active sand source

The *Pumice Vitrisols* are characterized by coarse-grained pumice (majority >2 mm in the top 10 cm of soil). Pumice is a porous material, which often floats on water. This porosity influences the hydrology, allowing for some water retention. These soils are common in the vicinity of Mt. Hekla (Fig. 6.15), along the active volcanic zone from the Torfajökull caldera to Vatnajökull glacier, and near the Askja caldera in NE highlands. There is a difference in soil properties depending on whether the pumice is rhyolitic (high SiO_2) or basaltic (low SiO_2) as more acidic soils tend to form in the silicic pumice. The basaltic pumice weathers more readily, which allows for ecosystem restoration, while the silicic pumice weathers very slowly, making restoration efforts more difficult. The pumice surfaces are often more stable than the sandy surfaces, but the lack of fine materials hampers natural succession after the volcanic events.

6.6 Other Soils

Other soil types than Andosols, Vitrisols and Histosols occur in Iceland. The fjords of western Iceland often have beaches that are calcareous from ocean derived sediments. These sediments are periodically blown inwards, affecting the neighboring soils, creating a transect from Andosols/Histosols towards Calcisols. These transects have not been studied.

Permafrost occurs in the highlands with some wetlands developing palsas with permanently frozen soil core, hence Cryosols (see Saemundsson et al. 2012). Permafrost is also likely to occur on the high deserts (e.g. >900 m elevations), but are poorly studied (see Chap. 10).

Bare rocks and gravelly scree slopes are classified as Leptosols, and they are common throughout the country.

6.7 The Mosaic

The purpose of this subchapter is to illustrate some Icelandic landscapes in photographs (Figs. 6.16, 6.17, 6.18, and 6.19), with emphasis on how variable the surfaces are. More than two soil types can occur within a range of a few meters. This makes mapping of the soils of Iceland often quite

Fig. 6.16 A typical valley landscape in Southeast Iceland (Lón). Much of the slopes are barren scree slopes, often of rhyolitic rocks associated with the interiors of Tertiary volcanic systems, exposed by the Quaternary glacial erosion. The ecosystems are degraded, with remnant soils at the bottom (Gleyic, Histic, and Brown Andosols), and in places shrublands that survived heavy land use over the ages, and is now spreading with reduced grazing and a shorter grazing period. Soil mapping capturing the details of the soil variability is difficult and involves grouping the soils into complexes of several soil types

Fig. 6.17 A valley in the Tertiary landscape of Skagi, North Iceland. Histosols in depressions (grasses and sedges), grading into Histic Andosols closer to dryland patches (heath dominating). The drylands have both Brown Andosols rich in organic matter (>6 % C in surface horizons), but some are classified as Histic Andosols (even in dryland positions). The soils are relatively thin (far from aeolian sources). The slopes have alternating Histic Andosols, Brown Andosols, and Leptosols (scree). Snowmelt is active in early summer

Fig. 6.18 Wetland landscape of West Iceland. The wetlands are Histosols of high resilience against land use pressures (see Chap. 12), but landuse-induced erosion has stripped the soil cover from the dryland parts, exposing Tertiary basaltic rocks. Some of the drylands have vegetation cover with Brown Andosols, including regrowth on formerly eroded land

Fig. 6.19 Sandy desert volcanic landscape in the southern highlands. The major glacial river Tungnaá in the background. Areas sheltered from the abrasion of the sand have vegetation cover and very sandy Brown Andosols, while the deserts are Sandy Vitrisols. This kind of landscape is characteristic of many of the southern and northeast highlands

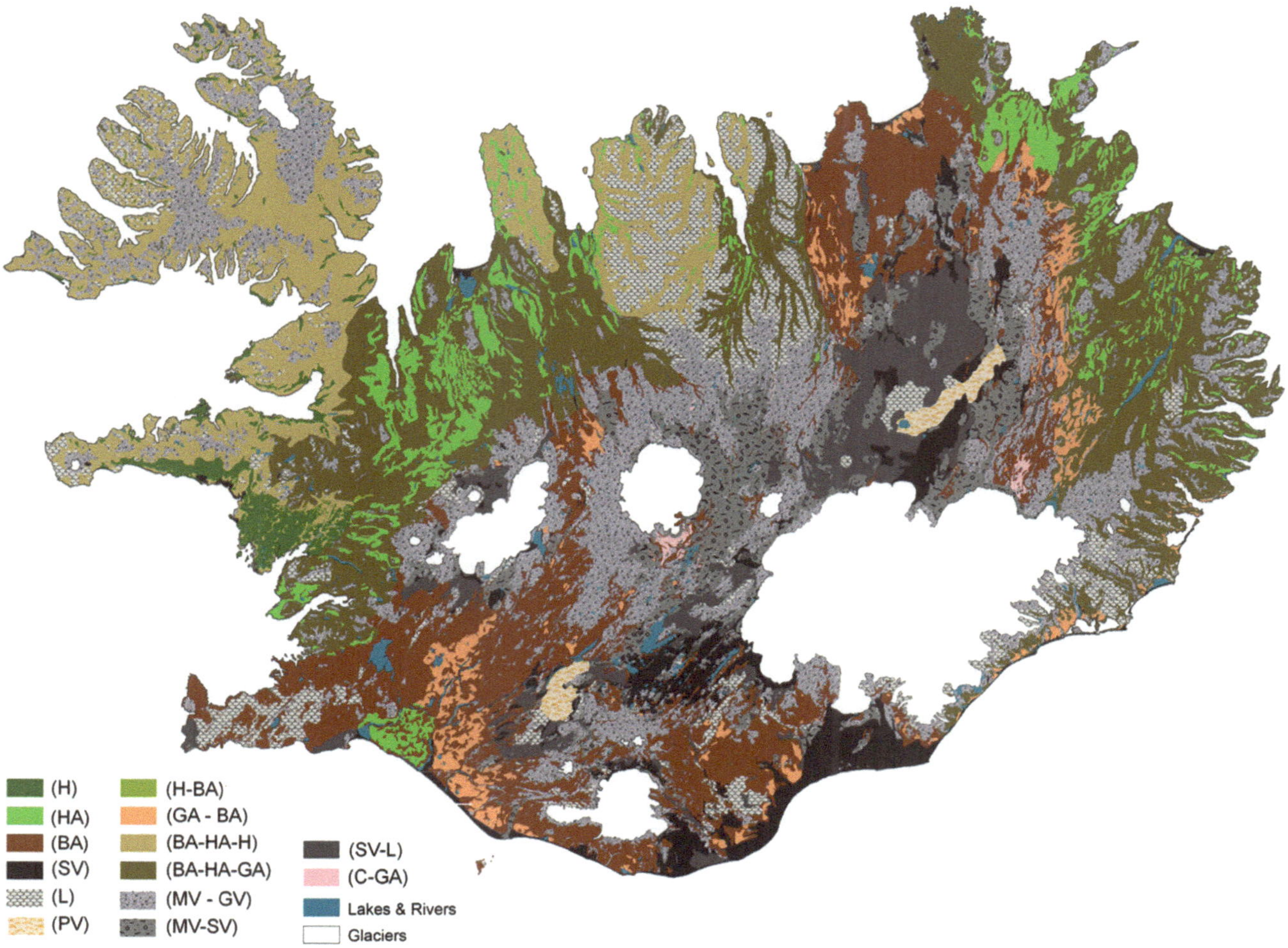

Fig. 6.20 A general soil map of Iceland from Arnalds and Oskarsson (2009). Note that most areas are defined as a complex of soil types, such as Brown Andosols and Gleyic Andosols alternating on the landscape

complicated, as one has to combine various soil types into "soil complexes" which are landscape units consisting of two or more soil types.

6.8 The Soil Map of Iceland

The first soil map of Iceland under the classification scheme presented here was first published digitally in 2001 but the current version (Fig. 6.20) is from 2009, published in *Náttúrufræðingurinn* (The Naturalist) by Arnalds and Oskarsson. It utilizes the AUI Soil Database and a large set of soil data accumulated for the IGLUD/LULUCF project also housed at the Agricultural University, based on random sampling of different vegetation types under variable geographic conditions. The Farmland Database (vegetation classes) is extensively used to delineate the mapping units verified by the AUI databases. The Soil Erosion Database is also extensively used for the desert soil types. This mapping procedure provided the basic data for Iceland to the *Soil Atlas of Europe* (Jones et al. 2005) and the *Soil Atlas of the Northern Circumpolar Region* (Jones et al. 2010). This current edition of the soil map is intended to give an overview of soils in Iceland at a relatively small scale (1:250,000). It is not intended to give accurate soil information for a given geographic point. It consists largely of soil complexes (see preceding subchapter), because of limitations given by this small scale. It is imperative to initiate work on the next level of the map, which would involve one step further down the classification, which has not been published to date.

References

Regarding punctuation and Icelandic characters in citations: See note on punctuation in the Preface

Arnalds O (1988) Soils of denuded areas in Iceland. Náttúrufræðingurinn 58:101–116 (in Icelandic, English summary)

Arnalds O (1990) Characterization and erosion of Andisols in Iceland. Unpublished PhD dissertation, Texas A&M University, College Station

Arnalds O (2004) Volcanic soils of Iceland. Catena 56:3–10

Arnalds O (2008) Soils of Iceland. Jökull—Icel J Earth Sci 58:409–421

Arnalds O (2010) Dust sources and deposition of aeolian materials in Iceland. Icel Agric Sci 23:3–21

Arnalds O, Kimble J (2001) Andisols of deserts in Iceland. Soil Sci Soc Am J 65:1778–1786

Arnalds O, Oskarsson H (2009) A soil map of Iceland. Náttúrufræðingurinn 78:107–121 (in Icelandic, English summary, table and figure legends)

Arnalds O, Hallmark CT, Wilding LP (1995) Andisols from four different regions of Iceland. Soil Sci Soc Am J 59:161–169

Arnalds O, Bartoli F, Buurman P, Oskarsson H, Stoops G, Garcia-Rodeja E (eds) (2007) Soils of volcanic regions in Europe. Springer, Heidelberg

Arnalds O,Thorarinsdottir EF, Thorsson J, Dagson Waldhauserova JP, Agustsdottir AM (2013) An extreme wind erosion event of the fresh Eyjafjallajökull 2010 volcanic ash. Sci Rep 3:1257. doi:10.1038/srep01257

Bartoli F, Burtin G (2007) Organo-mineral clay and physical properties in COST 622 European volcanic soils. In: Arnalds O, Bartoli F, Buurman P, Oskarsson H, Stoops G, Garcia-Rodeja E (eds) Soils of volcanic regions in Europe. Springer, Heidelberg, pp 469–491

Bjarnason OB (1952) Íslenskur mór (Icelandic peat). Fjölrit Rannsóknaráðs (Research Council report) no 3, Reykjavík

Bjarnason OB (1966) Íslenskur mór (Icelandic peat). Atvinnudeild Háskóla Íslands (University of Iceland), Reykjavik

Einarsson Th (1968) Geology, story of rocks and land. Mál og menning, Reykjavík (in Icelandic)

FAO-UNESCO (1988) Soil map of the world. Revised legend. World resources report 60. FAO, Rome

Gudmundsson Th (1978) Pedological studies of Icelandic peat soils. Unpublished PhD thesis, Department of Soil Science, University of Aberdeen, UK

Gudmundsson Th (1991) Soil in the revegetation area in the Blanda region. In: Thorsteinsson I (ed) Revegeation at the Audkulu and Eyvindarstadaheidi commons in North Iceland. Agricultural Research Institute, Reykjavik, pp 51–70 (in Icelandic)

Gudmundsson Th (1994) Jarðvegsflokkun FAO með hliðsjón af íslenskum aðstæðum (The FAO classification system adapted to Icelandic conditions). Rala report no 167. Agricultural Research Institute, Reykjavik (in Icelandic)

Gruner M (1912) Die Bodenkultur Islands. Archiv für Biontologie, Gesellschaft Naturforschender Freunde zu Berlin In: Komission bei R. Friedländer & Sohn

Hallsdottir M, Caseldine C (2005) The Holocene vegetation history of Iceland, state-of-the-art and future research. In: Caseldine C, Russell A, Hardardottir J, Knudsen O (eds) Iceland—modern processes and past environments. Developments in quaternary science vol 5. Elsevier, Amsterdam, pp 319–334

Hewitt AE (1993) Methods and rationale of the New Zealand soil classification. Landcare research science series no 1. Manaaki Whenau Press, Lincoln

Hewitt AE (1998) New Zealand soil classification. In: Landcare research science series no 2, 2nd edn. Manaaki Whenau Press, Lincoln

Hokkaido Soil Classification Committee (1979) Soil classification of agricultural land in Hokkaido 2nd approximation. Hokkaido National Agricultural Experiment Station Misc. Publ. 17, Sapporo, Japan

INRA (1998) A sound reference base for soils. Institute National de la Recherche Agronomique (INRA), París

IUSS Working Group WRB (2006) World reference base for soil resources. World soil resources reports no 103. FAO, Rome

Johannesson B (1960). The soils of Iceland. With a generalized soil map. Atvinnudeild Háskólans, Ritröð B nr 13. Reykjavík. Available through the Agricultural University of Iceland

Johannesson B (1988) Íslenskur jarðvegur. (The soils of Iceland). Reprinted version of a book from 1960 with extended appendix. Agricultural Research Institute (now Agricultural University of Iceland), Reykjavik

Jones A, Montanarella L, Jones R (eds) (2005) Soil Atlas of Europe. European Soil Bureau Network, European Commission, Office for Official Publications of the European Communities, Luxembourg

Jones A, Stolbovoy V, Tarnocai C, Broll G, Spaargaren O, Montanarella L (eds) (2010) Soil Atlas of the Northern Circumpolar Region. European Commission, Publications Office of the European Union, Luxembourg

Norddahl H, Ingolfsson O, Petursson HG, Hallsdottir M (2008) Late Weichselian and Holocene environmental history of Iceland. Jökull —Icel J Earth Sci 58:343–364

Nygard IJ, Johannesson B (1959) Soil map of Iceland (1:750,000). Adapted by the University Research Institute, Reykjavík (from original manuscript by Nygard IJ. Printed by the U.S. Geological Survey, included with the monographs by Johannesson; Soils of Iceland)

Olafsson JS (ed) (1998) Icelandic wetlands (Íslensk votlendi). Háskólaútgáfan (University Publishing), University of Iceland, Reykjavik, pp 121–129

Saemundsson Th, Arnalds O, Kneisel C, Jonsson HP, Decaulne A (2012) The Orravatnsrustir palsa site in Central Iceland—in an aeolian sedimentation environment. Geomorphology 167–168: 13–20

7 Physical Characteristics

7.1 Stratification of Soil Horizons

The aeolian and volcanic nature of soils of Iceland is strongly expressed by the clear horizonation of the soils, with abrupt horizon boundaries (Arnalds et al. 1995). This sometimes results in platy structures of the Brown and Gleyic Andosols, while granular structures are otherwise dominating in surface horizons of the Andosols. Subsurface horizons have a weakly developed blocky structure. However, it is often disrupted by tephra and aeolian layers within the volcanic zone (Fig. 7.1). The Vitrisols (soils of the deserts, see Sect. 6.5, Vitric Andosols according to WRB) are characterized by single grain structure, but very weakly developed blocky structure occurs in the Cambic Vitrisols. Pedon descriptions for nine characteristic pedons are presented at the end of the chapter, but chemical data for these pedons are presented at the end of Chap. 8. Some of these pedons are also used for explanations in Chap. 9.

The strong horizonation with alternating composition of parent materials, which consist mainly of basalts, but some andesites, and dacites, is much more pronounced than in other European volcanic soils (Taboada et al. 2007). Clear stratification representing changes in inorganic constituents is also evident in organic horizons. However, the effect of stratification becomes less evident with increasing distance from the active aeolian sources, with more homogenous mixing of the various aeolian deposits; once again emphasizing the importance of the aeolian environment and sedimentation for soil formation in Iceland.

The textural differences associated with lithological changes within the profile have important bearings on hydraulic characteristics. Coarse aeolian or tephra layers impede hydraulic conductivity when the soils are not saturated, hindering spread of water under unsaturated conditions into layers below (wetting process) or above (during drying out). Deep-rooted plants with roots growing through the coarse layers, such as birch trees, are better suited for such soils than shallow rooted plants.

7.2 Texture

Conventional methods for determining grain size distribution, such as the hydrometer method, have limited applicability for Andosols (e.g., Maeda et al. 1977; see Chap. 5). This is caused by strong aggregation of the clay constituents into silt-sized materials. Hand texturing, employing the broad categories of the textural triangle (USDA system), gives more reliable results than hydrometer methods for soils in Iceland. Figure 7.2 gives examples of clay contents measured by the hydrometer/pipette method and by ammonium oxalate extractions of the same samples. Allophane content is calculated as $Si_{ox} \times 6$ (Parfitt 1990) and ferrihydrite as $Fe_{ox} \times 1.7$ (Parfitt and Childs 1988). Total extracted clay is allophane and ferrihydrite combined. The graph clearly demonstrates that there is no relationship between the actual clay content (y-axis) and the clay content determined by the conventional pipette method. Buurman and van Doesburg (2007) showed that laser grain size analysis can be used to identify the sources of the aggregation in the soils from Iceland, but some of this aggregation is due to organomineral aggregation in addition to allophane clusters.

Particle size distribution from pedon descriptions are presented in Table 7.1. The table shows that soils in Iceland are most commonly silt loams, as was noted by Johannesson in 1960, with 40–47 % of the Andosol horizons textured as silt loams. Some horizons are textured as clay loams but none as clayey horizons (i.e., >30 % clay). The data show that Histic Andosols are finer textured on average than the Gleyic Andosols. Brown Andosols generally show more

O. Arnalds, *The Soils of Iceland*,
World Soils Book Series, DOI 10.1007/978-94-017-9621-7_7
© Springer Science+Business Media Dordrecht 2015

Fig. 7.1 Stratification of Brown Andosol within the volcanic zone. Tephra and aeolian sediments are frequently deposited on top, subsequently becoming buried layers that have abrupt horizon boundaries. Tephra layers, mostly from Mt. Hekla but also the Katla volcanic system, are prominent. Frost action creates wavy boundaries

variability in textures, with coarse textured horizons found in drylands close to volcanoes and aeolian sources. Many of the coarse textured horizons (loamy sand and sand) of the Andosols are tephra layers. The coarse textures reflect the young age of the soils, which are continuously buried under more recent aeolian and tephra sediments. With less aeolian deposition, more time is allowed for chemical weathering and clay formation. The texture of the Vitrisols is extremely variable, depending on the geology of each site, but sandy textures dominate.

The Andosols that form in the aeolian sediments are often devoid of coarse fragments, even profiles exceeding 1 m thicknesses, making the soil easy to work with, such as plowing for hay fields. However, there are some thin Andosol mantles overlying glacial till that tend to be quite gravelly due to frost heaving of rocks towards the surface.

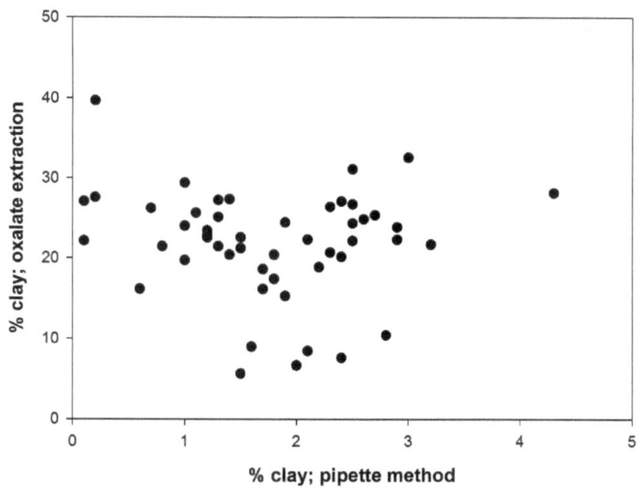

Fig. 7.2 Comparison of clay content determined by pipette method and oxalate extractions. Based on data for 50 horizons from four pedons located in different regions (from Arnalds 1990; see also Arnalds et al. 1995). The pipette method shows <5 % clay, while actual clay contents are mostly >10 %. There is no clear relationship, indicating that conventional texturing methods do not work for determining the clay fraction of these soils

7.3 Bulk Density

Icelandic Andosols are "light" or "fluffy" soils as is characteristic of Andosols. Low bulk density is one of the diagnostic criteria for Andosols (WRB and Soil Taxonomy), which is in part waived for vitric soils. Icelandic Andosols (soils under vegetation) generally have bulk density lower than 0.8 g cm^{-3}, but the Vitrisols (soils of the deserts) usually have bulk densities higher than 0.8 g cm^{-3}. Histic horizons with >20 % C generally have bulk density <0.4 g cm^{-3}. The low bulk density of the Andosols are clearly related to the carbon content of the soils overall (Fig. 7.3).

However, the range in bulk density for mineral horizons is quite variable as can be seen in Fig. 7.4 which shows bulk density data for horizons with 2–8 % carbon. There is a slight tendency for lower bulk density with increased carbon within this carbon range (left) and increased allophane (right). Total clay yields similar relationship with bulk density as the allophane shown on the right. These weak relationships show that other factors, such as density of the parent materials (often with numerous tephra layers (Fig. 7.5)) also influence bulk density at relatively low carbon levels, as many of these horizons are young with limited weathering.

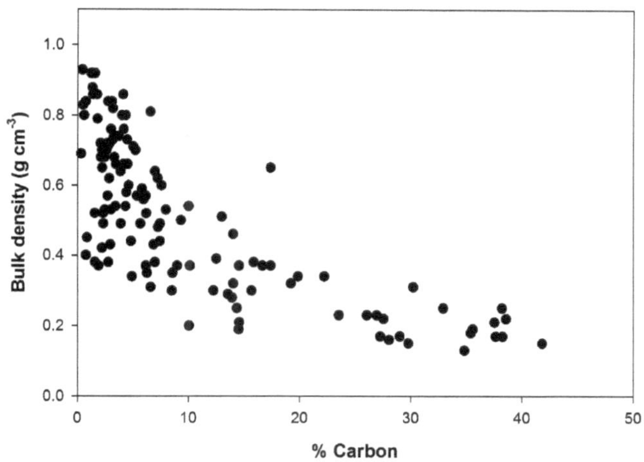

Fig. 7.3 Relationship between bulk density and carbon content for 160 horizons from various depths and soil types. Based on the AUI database, Arnalds (1990), Orradottir (2002), and Guicharnaud (2002). Vitric soil horizons with BD >1 excluded

7.4 Hydrological Characteristics

7.4.1 Infiltration

The fate of precipitation, when it reaches the surface, has a large impact on soil water relations and hydrology. Most of the summer precipitation infiltrates readily into soils in Iceland. However, a large part of the precipitation falls in wintertime when the ground is frozen, leading to surface runoff. Furthermore, parts of the winter precipitation falls as snow which is subjected to (i) large-scale removal due to snowdrifts, especially from higher ground and barren areas; (ii) periodic accumulation in depression areas and where taller vegetation is found (trees and shrubs); and (iii) rapid surface runoff from barren areas during snow melt because

Table 7.1 Grain size classes of the major soil types

Soil	Clay loam	Silt loam	Loam	Sandy loam	Loamy sand	Sand
	% within each soil type					
Histic Andosol	36	40	8	4	12[a]	0
Gleyic Andosol	17	44	19	10	10[a]	0
Brown Andosol	4	47	26	9	40[a]	3[a]
Vitrisol	2	6	2	38	44	8

Histosols are excluded. Histic Andosols are dominated by clay loam and silt loam; Gleyic and Brown Andosols are dominated by silt loam and loam. The Vitrisols are sandy and also tephra layers in the Andosols. Based on the AUI Database, about 200 horizons

[a] Mostly tephra layers

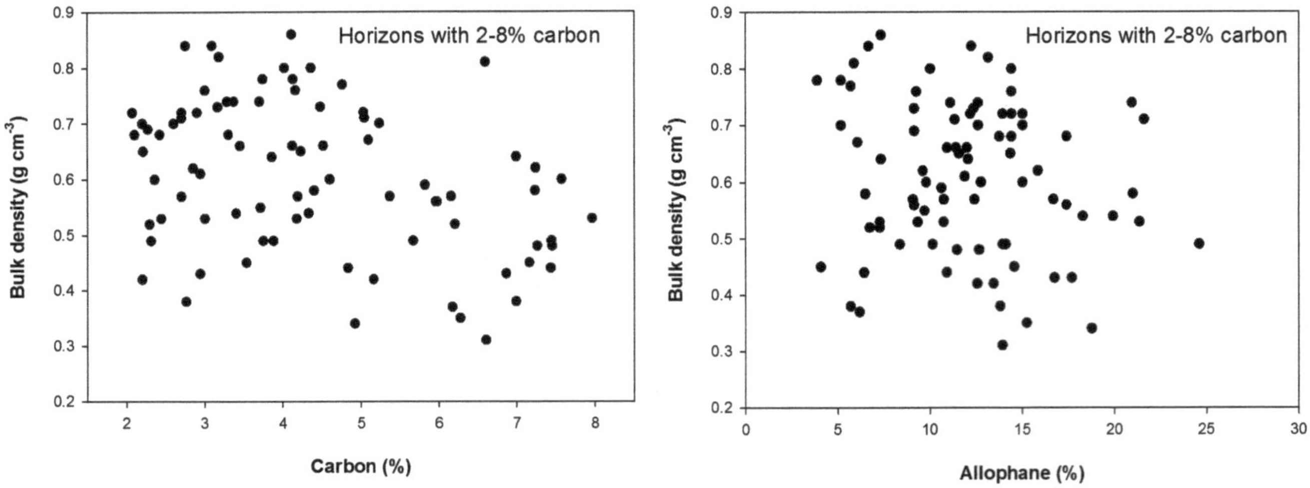

Fig. 7.4 Bulk density of 83 soil horizons with 2–8 % carbon as related tocarbon content (*left*) and allophane content (*right*). Based on AUI data

of limited infiltration during winter. Therefore, only a proportion of the precipitation infiltrates the soil, especially in desert areas and in the highlands. Furthermore, much of the precipitation seeps through the soil as gravitational water as the sandy Vitrisols have limited water holding capacity, enhancing water shortages in these systems, together with rapid evaporation of black (warm) surfaces during summer sunshine, as discussed in Chap. 4 about the deserts.

It is known that infiltration rates are generally high in Andosols, and also into the coarse-grained Vitrisols. Orradottir et al. (2006, 2008) investigated infiltration of various soils in Iceland and found that summer rates ranged between 28 and 363 mm h^{-1}. The summer rates are higher for sandy soils (102–363 mm h^{-1}) than finer textured soils (28–94 mm h^{-1}). These rapid rates indicate that most of the precipitation during summer will enter the soil and that water erosion is unlikely in most cases, except on steep slopes with loose Andosol remnants, also considering that high intensity rainstorms are not common under the sub-Arctic climate. Infiltration rates during winter, when the soils become frozen, are quite different from the summer rates. Winter rates are generally reduced in the Andosols (vegetated surfaces) except where vegetation cover is vigorous (e.g., birch forests), yet infiltration is allowed, but numbers vary greatly between vegetation and soil types (0–72 mm h^{-1}) with disturbed or heavily grazed sites having the lowest rates. Even where infiltration is greatly reduced due to frost, the vegetation cover reduces water erosion, but runoff is inevitable from the deserts in winter. Orradottir's research (Orradottir 2002; Orradottir et al. 2008) shows that in soils with limited vegetation cover (Vitrisols), a concrete type of ice can form that can impede infiltration completely. Higher winter infiltration rates sometimes found for sandy Vitrisols are most likely associated with relatively dry conditions, with little frost in the ground.

Impeded infiltration in winter has important consequences for the hydrology of the 40,000 km^2 of desert land in Iceland. Frequent snowmelt events in winter result in rapid surface runoff, with temporary flooding of rivers and loss of water from the desert ecosystems. Figure 7.6 shows a snowmelt event on frozen ground with coarse sand near Mt. Hekla, but surface water is only seen during such events.

Vegetation types have an interesting influence on the type of ice that forms in the soil during winter, with birch and rich grassland allowing for good infiltration (porous ice lenses), while planted coniferous patches and deserts have low winter infiltration rates (concrete ice; Orradottir et al. 2008). The research also indicates that heavy grazing reduces infiltration rates by causing soil compaction and altering vegetation vigor and the root mat, which in turn enhances the formation of concrete ice in the soil (Orradottir et al. 2006).

7.4.2 Water Retention

The water retention of the soils of Iceland varies considerably between soil types. Typical values are presented in Table 7.2. The Histosols and the Histic Andosols have extremely high water contents, even at the 15 bar (1.5 MPa) wilting point, and both Gleyic and Brown Andosols also have quite high water contents. The high water retention of the Icelandic Andosols is related to the nature of the porosity, which is usually characterized by capillary porosity, with limited macro porosity (Bartoli and Burtin 2007). The coarse-grained Brown and Gleyic Andosols of the volcanic zone generally have lower water retention than the fine-grained soils (30–50 % at saturation), which is exemplified in coarse Brown Andosols of the Gunnarsholt area (South Iceland) studied by Strachan et al. (1998).

Fig. 7.5 A soil profile being sampled for bulk density. Tin cans of known volume are driven into the different soil horizons. Profile from South Iceland with numerous tephra layers from the Hekla and Katla volcanoes. The coarse tephra layers have disruptive influence on water conductivity. AUI/OA/RAG

The numbers in Table 7.2 show considerable variability. This variability is demonstrated for 15 bar water retention in Fig. 7.7, where water retention in relation to clay and carbon content is plotted for low (<1.5 % C), median (1.5–8 % C) and high carbon content horizons (>8 % C). The water content is both related to the carbon contents (graphs to the right) and clay content (graphs on the left). This variability can be expected as these horizons are from soil profiles scattered around the country, taken from various depths. These graphs are based on horizons and do not represent soil types per se, as horizons within each profile most often have a range of clay contents, carbon values, and water characteristics, indicated by the wide scattering of the data.

At lower carbon contents, both clay and carbon contents contribute to the 15 bar water retention (upper four graphs in Fig. 7.7). However, at higher carbon contents, there is an

Fig. 7.6 Water on the surface during snowmelt on very sandy soils near Mt. Hekla. The ground has mostly concrete ice form with very slow infiltration rates. Water is never seen on the surface when the ground is not frozen (extremely high infiltration rates). *Photo* © Elin Fjola Thorarinsdottir (ISCS)

Table 7.2 Typical water holding capacities of Icelandic soil types

Soil type	Field capacity (0.3 bar)	Wilting point (15 bar)	Water holding capacity
		%	
Histosol	200–350	150–250	80–200
Histic Andosol	100–200	75–150	50–125
Gleyic Andosol	40–100	30–70	15–40
Brown Andosol	30–100	15–70	15–40
Vitrisol	5–40	5–30	5–15

Based on the AUI soil database. Data for sandy tephra layers are excluded

inverse relationship between allophane content and 15 bar water retention. This is because at high carbon contents, the pH becomes low and allophane formation is inhibited, but the water retention is governed by the increased carbon content (lowest graphs in Fig. 7.7).

Although the water retention values for Vitrisols (andic/vitric soils of barren areas) are considerably lower than for the other soil types, these soils do exhibit considerable water holding capacities that are much larger than in inorganic sandy soils found elsewhere. This is due to the inherent porosity of the vitric materials and the high water retention of the allophane present, even though it is in low quantities.

The high water retention of soils of Iceland has profound influence on the nature of Icelandic ecosystems. Given proper infiltration with good vegetation cover, these soils can store enormous water quantities, making the systems relatively resistant to drought conditions during the short growing season from May/June–August/September. This ability is, of course, affected by other properties, such as the presence of coarse tephra layers, which can block capillary rise of water within the pedon. Under such conditions deep roots become quite important for water transfer, both upward and downward, with deep-rooted trees such as mountain birch best suited for conditions close to the active volcanoes.

The Vitrisols and the sandy Brown and Gleyic Andosols do not have this water retention ability, and some of the Andosols are shallow, limiting the water storage capability. In addition, the dark surface of Vitrisols that lack vegetation cover warm up in sunlight, accelerating evaporation and soon resulting in drought conditions. Periodic droughts do occur in Iceland, in spite of the humid climate, typically lasting 3–6 weeks. This is enough time to cause severe water shortage in the dark surfaced Vitrisols, but also in sandy (vitric) Andosols. Plants in such areas usually have extensive root systems to increase water and nutrient availability.

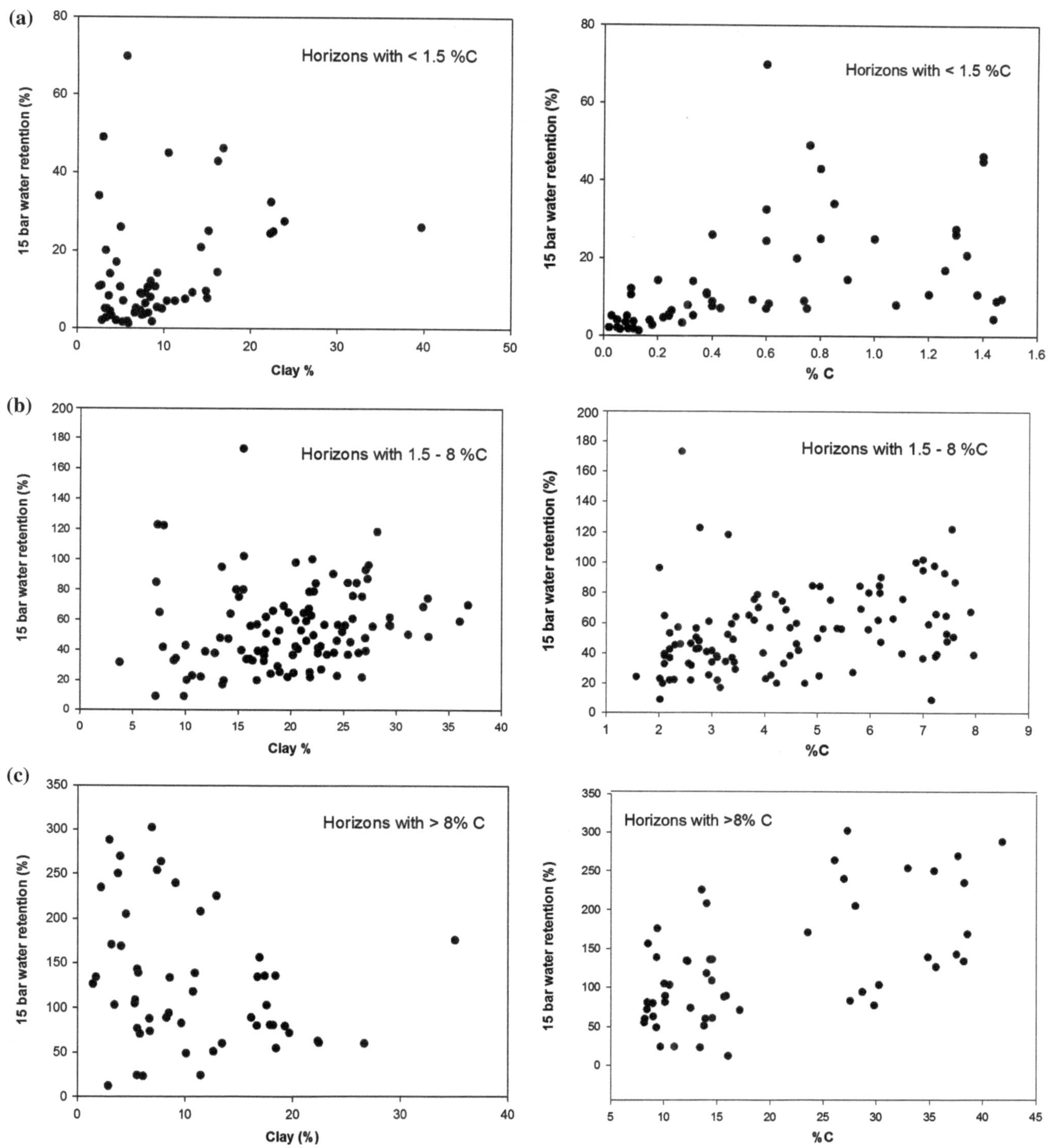

Fig. 7.7 15 bar (1.5 MPa) water content in soil horizons with **a** <1.5 % C, **b** 1.5–8 % C, and **c** >8 % C in relation to clay and carbon content. See explanation in text

7.5 Cohesion and Erosion Susceptibility

The high infiltration rates of soils in Iceland during frost-free periods reduce potential surface runoff, with minimal risk of water erosion except on sparsely vegetated slopes, but low water erosion risks are commonly reported worldwide for Andosols (McDaniel et al. 2012). However, water infiltration is greatly reduced in winter, leading to saturated soils over frozen subsurface layers. As Icelandic Andosols lack the phyllosilicates that provide other soil types with

Fig. 7.8 The noncohesive thixotropic nature and extremely high liquid limit of the Andosols is demonstrated by a clod of soil, and the same soil after applying some disturbance and pressure on the clod

cohesion, these saturated soils can be noncohesive. Soils in Iceland are often thixotropic in nature, meaning that they easily reach the liquid limit upon disturbance at high water contents, even though they seem cohesive before the disturbance (Fig. 7.8). In other words, the soils can contain large amounts of water and yet appear relatively dry until they are disturbed or become too saturated with water. When disturbed, the water is readily released. This can lead to massive soil erosion by water, and trigger landslides, which are common in Iceland (Fig. 7.9).

The unique water holding characteristics of Andosols, with their thixotropic nature is well illustrated by measurements of the plastic limits of the soils (Atterberg limits), and placing the soils on the well-known Casagrande plasticity chart. Most soils of the world fall somewhere near the A-line on the chart, but Andosols have a very narrow range from being plastic until they reach the liquid limit (low to zero plasticity index; PI). This occurs at very high water contents, placing the soils separate from other world soils near the zero PI line at high liquid limits (LL) as can be seen in Fig. 7.10.

Fig. 7.9 A massive landslide that occurred in the spring of 2013 in Northeast Iceland during spring thaw. Some parts of the surface were frozen, which allowed for building up high pressures in the soils from the snowmelt, resulting in shear failures. After the failure, the soils reach the liquid limit causing rapid flow of the materials down the slope

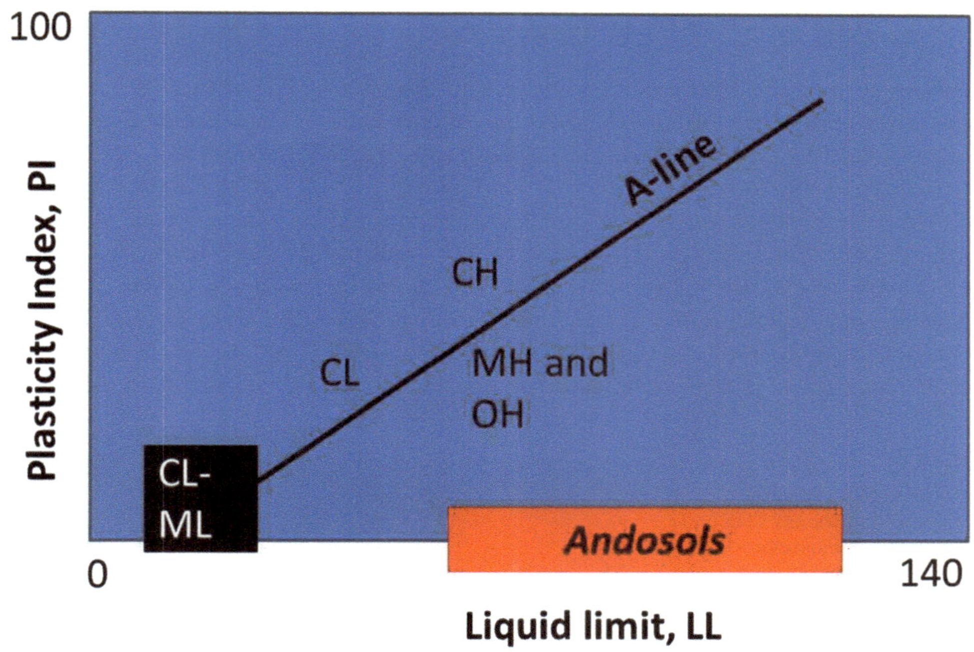

Fig. 7.10 Icelandic Andosols placed on the Casagrande plasticity chart. Based on Arnalds (1990)

The andic colloidal matter has a tendency to form silt-sized stable aggregates as is discussed in Chap. 5 on Andosols. Silt-sized particles generally lack cohesion, but are of the size easily picked up by the wind (saltation movement, see Chap. 11 on the aeolian environment). The thicker Icelandic Andosols (close to aeolian or volcanic sources) also lack coarse fragments, as the parent materials have been deposited by the wind or as volcanic ash. These soils are therefore extremely susceptible to wind erosionerosion, when the vegetation cover is disturbed. The wind and water are often "coworkers" at sites with disturbed Andosols in Iceland: wind erosion can be active during dry spells and water erosion during wet periods, especially in winter when infiltration rates are reduced (Fig. 7.11). This in part explains the extreme soil erosion that has occurred in Iceland over the last millennia (see Chap. 12).

Fig. 7.11 Bare soils with concrete ice haltering infiltration, exposed to water erosion. Severe wind erosion occurs at this site when the soils are dry. *Photo* © Berglind Orradóttir, AUI

7.6 Simplified Pedon Descriptions

Following are pedon descriptions for nine pedons of various characters obtained from a diverse range of environmental settings. Analytical data is presented for these soils in the subsequent chapters.

These descriptions come from various research campaigns: Three European COST-622 profiles, two from the Ph.D. work of the author, two desert profiles published by Arnalds and Kimble (2001), and finally two profiles obtained for the AUI soil database. The system used for each of the descriptions is kept unchanged. Site descriptions are shortened and the soil descriptions are slightly abbreviated. Photos of all the profiles (Figs. 7.12, 7.13, 7.14, 7.15, 7.16, 7.17, 7.18, 7.19, and 7.20) are also shown. *Note* Capital T means that the horizon is a single tephra layer, but tx has *x* number of well-noticeable tephra layers.

COST EUR-07; Ós, Northwest Iceland.

Described by AG Jongmans, F Van Oort and O Arnalds.

Soil: *Histic Andosol* (Iceland system), Orthidystri-Vitric Andosol (WRB 2001), Ashy, amorphic Eutric Pazchic Fulvicryand (Soil Taxonomy 1999). Frigid temperature regime, Udic moisture regime, MAT: 2.5 °C, MAP: 550 mm. Winters mild, summers cool.

Fig. 7.12 The Ós soil pedon (Northwest Iceland—Histic Andosol) from the EU COST-622 project. The hummocky landscape show well up in the profile (see pedon descriptions in this chapter and chemical data in Chap. 8)

Fig. 7.13 The Auðkúluheiði pedon (Northwest highlands—Brown Andosol) from the EU COST-622 project. Cryoturbated rhyolitic Hekla tephra layer prominent in the profile. (See pedon descriptions in this chapter and chemical data in Chap. 8)

Fig. 7.14 The Hella pedon (South Iceland—Gleyic Andosol) from the EU COST-622 project. Lower part of the soil is more organic (*darker*), but becomes *light colored* from *lighter colored* tephra fragments (aeolian deposition) and less organic matter after the Settlement. The "Settlement tephra layer" evident at the color change. *Upper part* of the profile cryoturbated (changes in landuse, vegetation cover and climate). (See descriptions in this chapter and chemical data in Chap. 8)

Gently rolling glaciated landscape with west facing gentle slope. The surface is hummocky with steep 20–50 cm high hummocks. The vegetation is "half-drained" grassland with vegetation strongly affected by grazing.

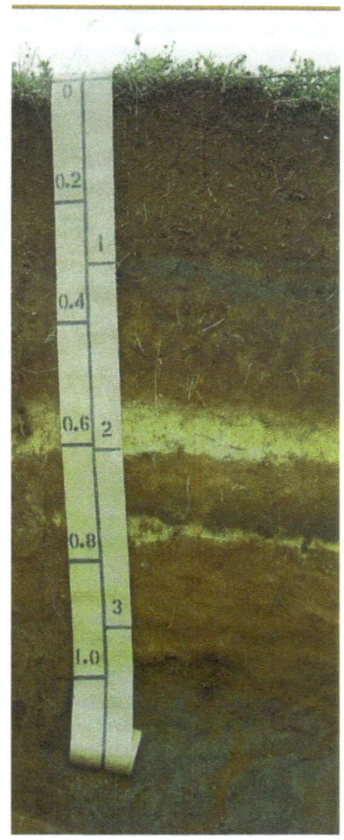

Fig. 7.15 The Godafoss Pedon (NE Iceland—Brown Andosol) from Arnalds (1990), Arnalds et al. (1995). Light colored tephra layers are from Mt. Hekla. Dark layer from about 30 cm is from 1480 AD. This pedon is discussed in further detail in Chap. 9 on genesis. (See descriptions in this chapter and chemical data in Chap. 8)

Bedrock is about 10,000 years glacial till, but about 1 m thick aeolian and tephra deposits on top, both basaltic and some rhyolitic grains, with distinctive tephra layers (Table 7.3; Fig. 7.12).

Pedon COST EUR-08; Auðkúluheiði. Northwest highlands.

Described by AG Jongmans, F Van Oort and O Arnalds.

Soil: *Brown Andosol* (Iceland system), Dystri-Vitric Andosol (WRB 2001), Ashy, amorphic Typic Vitricryand (Soil Taxonomy 1999). Cryic temperature regime, Udic moisture regime, MAT: 1.0 °C, MAP: 800 mm. Mild winter, frost 6–9 months, snow cover 3–7 months, summers cool.

Gently rolling glaciated landscape, nearly level. The surface is hummocky with steep 10–30 cm high hummocks. The vegetation is typical poor heathland, with *Betula nana*, *Empetrum nigrum* and *Vaccinium* spp. Vegetation strongly affected by grazing.

Fig. 7.16 The Thingvellir Pedon (SW Iceland—Brown Andosol) from Arnalds (1990), Arnalds et al. (1995). The soil has some relatively clay rich horizons (allophane + ferrihydrite > 35 %), but is lacking prominent tephra layers. (See descriptions in this chapter and chemical data in Chap. 8)

Fig. 7.17 The Möðruvellir Pedon (North Iceland—Histic Andosol) from the AUI Soil Database. The soil is >2 m thick with higher organic contents with depth. Light colored tephra layers are from Mt. Hekla. (See descriptions in this chapter and chemical data in Chap. 8)

Fig. 7.18 The Breiðdalur Pedon (Southern Eastfjords—Gleyic Andosol) from the AUI Soil Database. Dark tephra layer at 40 cm is from 1480 AD but the rhyolitc tephra from 1875 (Askja) and is disruptive to water conductivity in places. The *upper part* is quite cryoturbated. *Darker upper color* is in part because of increased aeolian deposition of black basaltic tephra grains from the growing desert in Northeast Iceland

Fig. 7.19 The Sigalda Pedon (South highlands—Cambic Vitrisol) from Arnalds and Kimble (2001). The soil is sandy but some allophane and organic matter are found in the Bw horizon

Bedrock is 9,000–10,000 years old glacial till, but 50–100 cm thick aeolian and tephra deposits on top, both basaltic and some rhyolitic grains, with distinctive tephra layers (Table 7.4).

Pedon COST EUR-09; Hella. South lowlands.

Described by AG Jongmans, F Van Oort and O Arnalds.

Soil: *Gleyic Andosol* (Iceland system), Thaotihisti-Vitric Andosol (Umbric and Pachic) (WRB 2001), Ashy, amorphic Eutric Pachic Fulvicryand (Soil Taxonomy 1999). Frigid temperature regime, Udic moisture regime, MAT: 4.5 °C, MAP: 1,150 mm. Summers cool, mild winters, 4–5 months growing season. Plain of southern lowlands. The surface is hummocky. The vegetation type is *grassland*. Very noticeable environmental change about the Settlement 1,125 years ago, increased aeolian deposition resulted in higher mineral content in the sediments and lighter soil colors. Profile is situated in a 2-year-old ditch, 2.5 m deep. Parent materials

Fig. 7.20 The Mýrdalssandur Pedon (South lowlands—Sandy Vitrisol) from Arnalds and Kimble (2001). The soil has formed in recent (1918) glaciofluvial deposits that are continuously affected by erosion losses and aeolian additions

are aeolian sediments, many tephra layers, and organic materials. Site is near Mt. Hekla and is influenced by volcanic activity in both Mt. Hekla and Katla (Table 7.5).

Pedon Godafoss, Northeast Iceland.

Described by O Arnalds, CT Hallmark and LP Wilding.

Soil: *Brown Andosol* (Icelandic scheme), ashy over medial Vitric Haplocryand (Soil Taxonomy).

Soil formed in aeolian and tephra materials ("eolian-andic") overlying about 9,000 year old glacial till. Summit position, gently rolling, 260 m a.s.l. Well drained. Land is open range for sheep grazing with typical heath vegetation (poor heathland). Source: Arnalds (1990), Arnalds et al. (1995) (Table 7.6).

Pedon Thingvallasveit. Southwest Iceland.

Described and sampled by LP Wilding, CT Hallmark and O Arnalds.

Soil: *Brown Andosol* (Icelandic scheme); medial Typic Haplocryand (Soil Taxonomy).

Soil formed in volcanic aeolian materials ("aeolian-andic") overlying about 9,000 year old glacial till. About 200 m elevation on gently rolling glacial till, 3° slope SSE. Moderately drained site with moss heath vegetation (poor heathland) with *Rachomitrium* moss together with common dwarf shrub heath such as *Calluna*, *Empetrum* and *Vaccinium* spp. Source: Arnalds (1990), Arnalds et al. (1995) (Table 7.7).

PEDON Möðruvellir II. North Iceland.

Described and sampled by Arnalds et al. (2001).

Soil: *Histic Andosol* (Icelandic scheme).

Profile taken at side of newly restored manmade ditch at 24 m elevation at slightly sloping site. Site used for hay production. Parent materials made of different mixture of aeolian materials and organic residues, interrupted by few major tephra layers (Hekla tephra). Soil qualifies as peat (>20 % C) in many subsurface horizons, but not in the control section of the surface (Table 7.8).

Pedon Breiðdalur. Southeast Iceland.

Described and sampled by O Arnalds, R Guicharnaud and B Oladottir.

Soil: *Gleyic Andosol* (Icelandic scheme).

Profile dug into grassland/rich heathland in a valley bottom at 130 m elevation. Grass species and dwarf shrub heath vegetation. Rather poorly drained site. Rare confirmed occurrence of smectite at the bottom of the profile (Table 7.9).

Pedon Sigalda. Southern highlands.

Described and sampled by Olafur Arnalds and Asa L. Aradottir.

Soil: *Cambic Vitrisol* (Icelandic scheme); ashy, shallow, amorphic Typic Vitricryand (Soil Taxonomy)

The site is sandy undulating glacial till influenced by aeolian processes and frost heaving of coarse fragments. *Desert surface*; there is almost no vegetation cover (0.5 %). The little vegetation there is remains grazed. Elevation is 540 m a.s.l., MAP 1,200 mm, average July mean temperature about 7 and −5 °C in January. Source: Arnalds and Kimble (2001) (Table 7.10).

Pedon Myrdalssandur, south lowlands, glacial floodplain.

Described and sampled by Olafur Arnalds and Asa L. Aradottir.

Soil: *Sandy Vitrisol* (Icelandic scheme); ashy, shallow, amorphic Typic Vitricryand (Soil Taxonomy).

Unstable glaciofluvial floodplain, most of current surface formed during massive "jökulhlaup" flood during a Katla eruption. Active aeolian processes. Materials consisting of basaltic glass and pumice and some more crystalline fragments. Site at about 30 m elevation, MAP about 1,800 mm, average July temp about 10 °C and about 0 °C in January. The surface is sandy *desert* with about 1 % vegetation cover. Source: Arnalds and Kimble (2001) (Table 7.11).

Table 7.3 Pedon COST EUR-07, Ós. NW Iceland, Fig. 7.12

Horizon	Depth (cm)	Munsell color	Description
O	0–5	5YR 3/2	Dominantly partly decomposed organic material; some mineral particles are present; very weak fine granular structure; clear and smooth to:
Ah	5–17	5YR 3/3	Loam; weak, fine subangular blocky/granular structure; friable; pores are not detectable; common very fine and fine roots; abrupt and wavy to:
AC	17–35/50	2.5YR 3/4	Stratified horizon; organic clay loam; moderate fine platy structure; friable; few fine to medium iron mottles, brown to dark brown (7.5YR 4/4); common, fine to coarse roots; smooth to:
2BC	35/50–65	5YR 3/2 and 5YR 2.5/2	Organic loam; no macro structure; friable; common fine and medium roots; few, distinct iron mottles, red (2.5YR 5/8); smooth to:
3BC	65–73	10YR 6/6 and 10YR 5/4	Organic loam; no macro structure; friable; common fine and medium roots; few, distinct iron mottles, red (2.5YR 5/8); smooth to:
3CB T	73–82	10YR 5/4 and 5YR 2.5/1	Stratified horizon, consisting of H4 tephras from Hekla (4,000 years BP) having two colors: lower part yellowish brown (10YR 5/4), upper part (3 cm) very dark gray to black (5YR 2.5/1); thickness of the individual layers range from 0.5 to 4 cm; light colored tephra: loam; dark colored tephra: organic clay loam; moderate fine platy structure; friable; common fine and medium and few coarse roots; very few, fine and medium iron mottles, red (2.5YR 5/8); abrupt and smooth to:
4Bw	82–90/100	5YR 3/3	Organic clay loam; few coarse partly altered basaltic gravel and stones; weak, fine subangular blocky; friable; common fine roots; few, medium, distinct iron mottles, red (2.5YR 5/8), dominantly concentrated along pores and stones; abrupt and wavy to:
4B/Cg	>90/100	5Y 4/1	Clay; many coarse gravel to boulders, subangular, altered; no macro structure; friable; common, medium and coarse, prominent iron mottles, red (2.5YR 4/8), dominantly concentrated around pores and stones; few fine roots

Table 7.4 Pedon COST EUR-08, Auðkúluheiði, NW highlands, Fig. 7.13

Horizon	Depth (cm)	Munsell color	Description
O	0–3	–	Partially decomposed organic materials; abrupt and smooth to:
Ah1	3–11/19	7.5YR 3/3	Sandy loam; moderate fine to medium platy; friable; many fine to medium roots; from 12 to 16 cm irregular bright dark yellowish brown (10YR 4/5) spots, 1 cm; abrupt and wavy to:
Ah2	11/19–21/27	5YR 3/3	Sandy loam; no macro structure; very friable; no biopores detectable; common fine roots; abrupt and tonguing to:
Bw1	21/27–26/34	7.5YR 4/4	Sandy loam; no macro structure; very friable; few fine pores; common fine roots; abrupt and tonguing to:
2Bw2 T	26/34–37/42	10YR 5/5	Bright discontinuous tephra layer, slightly weathered; sandy loam; moderate fine platy structure; friable; common fine roots; occurrence of cryoturbationCryoturbation features, presence of material from the over- and underlaying horizons; abrupt and wavy to:
3Bw3/4C	37/42–59/62	7.5YR 3/5	30 % 4C, glacial till material, dark yellowish brown (10YR 4/4); loam; few fine gravel, subrounded, up to 3 cm, slightly weathered; no macro structure; friable; few, fine roots; abrupt and smooth to:
4C	>59/62	10YR 4/4	Glacial till; sandy clay loam; fine gravel to boulders, subrounded, partly weathered; layers of 2 cm consisting of 3Bw material, distributed inclined to the soil surface as a result of cryoturbation; no macro structure; friable; few fine roots in the upper 10 cm

Table 7.5 Pedon COST EUR-09, Hella, S-Iceland, Fig. 7.14

Horizon	Depth (cm)	Munsell color	Description
Ah	0–55	5YR 3/3	Loam; the horizon has a high organic matter content; locally pockets of angular, fine gravely and sandy material; disturbed stratification, the entire horizon shows features of cryoturbation
2C T	55–60	2.5YR 2.5/0	A basaltic ash layer of 5 cm, very dark gray to black with a loamy sand texture; friable; common fine to medium roots; clear and wavy to:
3H	60–95	7.5YR 3/3	Stratified horizon consisting of plant remnants; friable; common fine and medium roots; occurrence of discontinuous fine bands of basaltic ash; clear and wavy to:
3C	95–100	10YR 3/2	95–100 cm; very dark grayish brown (10YR 3/2), with light spots of sand grains, light yellowish brown (10YR 6/4); no macro structure; friable; few fine roots; abrupt and wavy to:
4H	100–230	2.5YR 2/0	Stratified horizon consisting of organic material with aeolian mineral and basaltic pyroclastic [black (2.5YR 2/0)] layers with a loamy sand texture and ranging in thickness from 1 to 10 cm; at 160 cm some mineral layers up to 2 cm consisting of lighter colored rhyolitic material; organic material and ash layers are friable; at 120 and 170 cm wood remnants of birch up to 20–30 cm in size

Table 7.6 Pedon Godafoss, N-Iceland, Fig. 7.15

Horizon	Depth (cm)	Munsell color	Description
A1	0–4	7.5YR 3/2	Mucky loam; weak fine granular parting to weak fine platy structure; very friable; many fine roots; clear smooth boundary
A2	4–12	5YR 3/3	Mucky loam; weak fine subangular blocky parting to weak fine granular structure; very friable; many fine roots; clear smooth boundary
A3	12–20	5YR 3/3	Loam; weak fine subangular blocky parting to weak fine granular structure; very friable; many fine roots; clear smooth boundary
A4-(t1)	20–26	7.5YR 3/2	Loam; weak fine subangular blocky parting to weak fine granular structure; very friable; many fine roots; includes thin black (10YR 2/1) tephra layer; abrupt wavy boundary
Bw1-T	26–29	10YR 2/1	Loamy fine sand; weak medium subangular blocky structure; very friable; common fine roots; tephra layer AD 1480; abrupt wavy boundary
Bw2-(t1)	29–41	5YR 3/2	Silt loam; weak medium and coarse subangular blocky structure; friable; common fine roots: pockets of a remnant yellowish red (5YR 5/6) tephra layer (H1?) incorporated in horizon, about 30 % (by volume) of horizon; clear smooth boundary
Bw3	41–49	5YR 3/3	Loam; weak medium and coarse subangular blocky structure; friable; common fine roots; abrupt wavy boundary
Bw4-T	49–57	10YR 5/4	Silt loam; many medium distinct strong brown (7.5YR 5/6) mottles; weak medium platy parting to weak medium subangular blocky structure; friable; common fine roots; many fine vesicular pores; horizon is H3 tephra (2900 BP); abrupt wavy boundary
Bw5	57–65	5YR 4/4	Silt loam; weak medium subangular blocky parting to medium granular structure; friable; common fine roots; about 10 % (by volume) of horizon contains dark reddish brown (5YR 3/3) material from horizon below; clear smooth boundary
Bw6	65–70	5YR 3/2	Loam; weak medium subangular blocky structure; friable; few fine roots; lower boundary is irregular on microscale with lobes of the underlying white tephra extending abruptly into the horizon; pockets are slightly brittle; abrupt wavy boundary
Bw7-T	70–73	10YR 6/4	Silt loam; many fine distinct strong brown (7.5YR 5/6) mottles; weak medium palty structure; friable; few fine roots; horizon is H4 tephra (4000 BP); mottles occur along plate surfaces; root channels and vesicular pores; abrupt wavy boundary
Bw8-(t1)	73–91	5YR 4/6	Silt loam; weak medium and coarse subangular blocky structure; friable; common fine roots; dark reddish brown (5YR 3/3) pockets of loam-textured material are mixed in the horizon; a thin 1–2 cm brown (10YR 5/3) tephra layer occurs in the horizon; many fine vesicular pores; clear wavy boundary
2Bw9	91–101	5YR 3/4	Loam; weak coarse subangular blocky structure; slightly brittle; firm; few fine roots; coarse fragments are gravel-sized and originate from below; lower portion of horizon is olive brown (2.5Y 4/4) grading to the till below; 10 % coarse fragments; clear wavy boundary
2C	101–121	5Y 4/2	Gravelly loam; weak coarse platy structure; friable; few fine roots; glacial till is relatively dense as fines are packed between coarse fragments (subrounded basalt gravel and cobbles); about 10 % of horizon is gravels >2 cm; a few oxidized dark reddish brown (5YR 3/3) vertical planes extend through horizon; 25 % coarse fragments

Table 7.7 Pedon Thingvallasveit, SW Iceland, Fig. 7.16

Horizon	Depth (cm)	Munsell color	Description
A1	0–12	7.5YR 3/2	Silt loam; weak very fine granular structure; very friable; many fine pores; many fine roots; clear smooth boundary
A2	12–28	5YR 3/3	Silt loam; weak medium subangular blocky parting to weak fine granular structure; very friable; many fine pores; many fine roots; clear smooth boundary
Bw1	28–61	5YR 3/4	Silt loam; weak coarse subangular blocky parting to weak very coarse platy structure; very friable; many fine pores; common fine roots; clear smooth boundary
Bw2	61–68	10YR 4/3	Silt loam; weak coarse subangular blocky parting to moderate coarse platy structure; friable; many fine pores; common fine roots; abrupt smooth boundary
2Bw3	68–87	10YR 4/2	Silt loam; few medium faint dark yellowish brown (10YR 4/4) mottles; moderate very coarse platy structure; friable; few fine roots; occasional vertical fractures noted; abrupt smooth boundary
2C	87–142	10YR 4/1	Silt loam; few medium distinct yellowish brown (10YR 5/4) and common medium distinct brown (7.5YR 4/4) mottles; moderate medium platy parting to strong fine platy structure; firm; very few fine roots; structure is inherited from parent material; mottles along root channels and some bedding planes; very occasional vertical planes

Table 7.8 Pedon Mödruvellir II, N Iceland, Fig. 7.17

Horizon	Depth (cm)	Munsell color	Texture and structure
A1 (H)	0–30	7.5 YR 5/6	Loam. Very weak fine granular structure. Very friable. Many very fine to medium roots. No mottles. Abrupt wavy boundary to:
2O1	30–55	10 YR 4/3	Very weak thin platy structure. Friable. Many very fine and fine roots. No mottles. Abrupt wavy boundary to:
2O2	55–83	5 YR 3/2	Weak fine and medium subangular blocky structure. Friable. Many very fine and fine roots. No mottles. Stratification of organic matter and aeolian matter expressed by different colors. Clear wavy boundary to:
2O3	83–98	5 YR 2.5/2	Very weak thin platy and very weak fine and medium subangular blocky structure. Friable. Mixed with yellowish undecomposed organic matter. No mottles. Abrupt wavy boundary to:
3C - T	98–104	10 YR 7/4	Sandy loam. Very weak subangular blocky structure. Firm. Very few fine roots. Few faint mottles. Tephra layer H3 (Hekla 2800 BP). Clear wavy boundary to:
4O1	104–145	10 YR 2.5/1	Very weak thin platy and weak fine subangular blocky structure. Few very fine and fine roots. Tephra layer in bottom, 1–2 cm, 10 YR 6/4, light reddish brown. Abrupt wavy boundary to:
4O2	145–180	5 YR 2.5/1	Weak thin platy structure. Few very fine and fine roots. Abrupt wavy boundary to: Weak thin platy structure
4O3	180–200	5 YR 2.5/1 & 5 YR 4/4	Weak thin platy and weak medium subangular blocky structure. Very few very fine and fine roots. Faint reddish mottles around roots (oxidized). Abrupt wavy boundary to:
4O4	200–260	5 YR 2.5/1	Weak thin platy and weak medium subangular blocky. Very few very fine roots. Up to 2 diameter centimeter stems of woody plants. Clear wavy boundary to:
4O5	260–300	5 YR 2.5/1	Very weak thin platy and very weak medium subangular blocky structure. Very few very fine roots. Up to 6 cm stems of woody plants. No mottles. Abrupt wavy boundary to:
4O6	300–350+	5 YR 2.5/1	Very weak thin platy and very weak medium subangular blocky structure

Table 7.9 Pedon Breiðdalur, Eastfjords, Fig. 7.18

Horizon	Depth (cm)	Munsell color	Texture and structure
A1	0–11	5YR 3/2	Silt loam; weak, medium granular structure; very friable; abundant roots of all sizes clear wavy boundary
A2-t	11–26	5YR 3/2	Silt loam; very weak, medium subangular blocky and very weak, medium granular structure; very friable; many fine and very fine roots; cryoturbationCryoturbation signs; tephra (possibly "a" tephra from 1480) at the bottom, 1–3 cm thick; very abrupt wavy boundary
Bw1-t	26–37	7.5YR 3/4	Silt loam; weak, medium subangular blocky structure; many fine and very fine roots; 2 % medium, cylindrical mottles; light colored 0–1 cm tephra at bottom (Öræfajökull 1362?); cryoturbation signs; very abrupt, wavy boundary
Bw2	37–47	10YR 3/2	Silt loam; weak subangular blocky structure; very friable; moderately few, medium roots; dark yellowish brown (10 YR 4/6) clear cylindrical mottles (1–3 mm); very abrupt wavy boundary
Bw3	47–70	10YR 3/4	Silty clay loam; weak, medium subangular blocky structure; friable; very few (1 %), medium mottles; few fine and very fine roots; cryoturbation signs; very abrupt wavy boundary
Bw4	70–82	10YR 3/3	Silty clay loam; weak, medium subangular blocky structure; friable; few, fine and very fine roots; very few (4 %) medium mottles; very abrupt wavy boundary
2Bw1	82–100	2.5 Y 3/2 2.5Y 6/4 2.5YR 4/8	Gravelly clay; weak, fine granular structure; friable; few, fine and very fine roots; clay caps on gravel fragments; many (30 %) mottles; rounded gravel 70 %; very abrupt wavy boundary
2Bw2	120–120+	10YR 4/1	Clay loam; medium subangular blocky structure; friable; very few fine roots; rounded 2–20 cm gravel; 5 % red and yellow mottles

Table 7.10 Pdeon Sigalda, South-Cenral, Fig. 7.19

Horizon	Depth (cm)	Munsell color	Description
C	0–2	NA	Gravel, wide range of diameters
2A1	2–22	10YR 3/2	Sandy loam; weak fine and medium granular structure; very friable; abrupt wavy boundary
2A2	22–34	10YR 3/2	Sandy loam; weak fine and medium granular structure; very friable; clear wavy boundary
3C	34–50+	2.5YR 2.5/0	Loamy sand; single grain

Table 7.11 Pedon Myrdalssandur, south lowlands, Fig. 7.20

Horizon	Depth (cm)	Munsell color	Description
A	0–4	7.5YR 2.5/0	Gravelly loamy sand; weak fine granular structure / structureless; very friable; abrupt wavy boundary
C1	4–8	7.5YR 2.5/0	Gravelly loamy sand; single grain; abrupt smooth boundary
C2	8–13	7.5YR 2.5/0	Loamy sand; single grain; abrupt wavy boundary
C3	13–25+	7.5YR 2.5/0	Gravelly sand; single grain

References

Regarding punctuation and Icelandic characters in citations: See note on punctuation in the Preface

Arnalds O (1990) Characterization and erosion of Andisols in Iceland. Unpublished PhD dissertation, Texas A&M University, College Station, Texas

Arnalds O, Kimble J (2001) Andisols of deserts in Iceland. Soil Sci Soc Am J 65:1778–1786

Arnalds O, Hallmark CT, Wilding LP (1995) Andisols from four different regions of Iceland. Soil Sci Soc Am J 59:161–169

Bartoli F, Burtin G (2007) Organo-mineral clay and physical properties in COST-622 European volcanic soils. In: Arnalds O, Bartoli F, Buurman P, Oskarsson H, Stoops G, Garcia-Rodeja E (eds) Soils of volcanic regions in Europe. Springer, Heidelberg, pp 469–491

Buurman P, van Doesburg JDJ (2007) Laser-diffraction grain-size analyses of reference profiles. In: Arnalds O, Bartoli F, Buurman P, Oskarsson H, Stoops G, Garcia-Rodeja E (eds) Soils of volcanic regions in Europe. Springer, Heidelberg, Germany, pp 453–468

Guicharnaud R (2002) Rúmþyngd í íslenskum jarðvegi (Bulk density of Icelandic soils). Unpublished B.S. thesis, University of Iceland, Reykjavík

Johannesson B (1960) The soils of Iceland. With a generalized soil map. Atvinnudeild Háskólans, Ritröð B nr 13. Reykjavík. Available through the Agricultural University of Iceland

Maeda T, Takenaka H, Warkentin BP (1977) Physical properties of allophane soils. Adv Agron 29:229–264

McDaniel PA, Lowe DJ, Arnalds O, Ping C-L (2012) Andisols. In: Huang PM, Yuncong L, Sumner ME (eds) Handbook of soil sciences. Properties and processes, 2nd edn. CRC Press, Boca Raton, pp 33-29–33-48

Orradottir B (2002) The influence of vegetation on frost dynamics, infiltration rate and surface stability in Icelandic rangelands. M.Sc. thesis, Texas A&M University, College Station, Texas

Orradottir B, Arnalds O, Thorsson J (2006) Ísig vatns í jarðveg: Áhrif gróðurs og frosts (Infiltration: the influence of vegetation and frost). Fræðaþing landbúnaðarins (Agricultural Science Congress) 2006:102–107 (in Icelandic)

Orradottir B, Archer SR, Arnalds O, Wilding LP, Thurow TL (2008) Infiltration of icelandic Andisols: the role of vegetation and soil frost. Arct Antarct Alp Res 40:412–421

Parfitt RL (1990) Allophane in New Zealand—a review. Aust J Soil Res 28:343–360

Parfitt RL, Childs CW (1988) Estimation of forms of Fe and Al: a review, and analysis of contrasting soils by dissolution and Moessbauer methods. Aust J Soil Res 26:121–144

Strachan IB, Arnalds O, Palmason F, Thorgeirsson H, Sigurdsson BD, Sigurdardottir H, Novoselac G (1998) Soils of the Gunnarsholt experimental plantation. Icel Agric Sci 12:27–38

Taboada T, Garcia C, Martinez-Cortizas A, Novoa JC, Potevedra X, Garcia-Rodeja E (2007) Chemical of reference European volcanic soils. In: Arnalds O, Bartoli F, Buurman P, Oskarsson H, Stoops G, Garcia-Rodeja E (eds) Soils of volcanic regions in Europe. Springer, Heidelberg, pp 307–323

8 Chemical Characteristics

8.1 Introduction and pH

Most soils of Iceland show chemical characteristics which are typical of Andosols, yet many of these soils are youthful, especially those close to aeolian and volcanic ash sources. The northerly location and cold climate are also revealed in the soil by accumulation of organic matter and slow decomposition. The volcanic and climatic influences are reflected by the chemical properties, such as pH and charge properties. These general characteristics are described here, but a table, showing common chemical properties for nine pedons with a total of 66 soil horizons, is presented at the end of the chapter. This chapter draws from many sources. The Agricultural University of Iceland (AUI) soil database is a prominent source of information, together with data from European COST-622 research cooperation on volcanic soils of Europe. Information is also drawn from various studies of the author and other sources, many of which were originally published in Icelandic.

The pH of soils in Iceland has a narrower range than often experienced in other countries of more diverse geology, because of the dominant influence of the aeolian materials, recharging the soils with basic cations. Furthermore, the values are higher than commonly reported for Andosols elsewhere in the world, including Alaskan Andosols (Ping et al. 1988, 1989; see also Arnalds et al. 1995). However, there are regional differences that are depicted for surface horizons of Brown Andosols in Fig. 8.1 (pH measured in water). The range is generally between 5 and 7 for the Brown Andosols, with the highest pH found in Northeast Iceland, but the lowest in the west and in some areas of the Westfjords. The northeast receives high amounts of aeolian additions and is also the driest part of the country, which minimizes lowering of the pH by leaching. Southeast and South Iceland receive the largest amount of precipitation, but the leaching is balanced by high aeolian additions. West Iceland has relatively lower aeolian additions and moderate rainfall, resulting in relatively low pH values. The pH of soils of the Westfjords is variable, ranging from 4.5 to 6.5 for Brown Andosols. This overview image shows similar results as was presented in a map by Johannesson (1960), which also suggested that pH was controlled in part by the aeolian additions and rainfall patterns.

The pH of wetland soils (Gleyic and Histic Andosols and Histosols) is usually 0.5–1 unit lower than for the Brown Andosols. The lowest pH is measured in Histosols in the Westfjord area, with pH 4 in the surface horizons, and even lower values found in subsurface horizons. Similar values were found in some of Histosols in West Iceland (Borgarfjörður and Mýrar).

The soil reaction of the Vitrisols of the deserts (Vitricryands according to ST) is generally higher than 7, often around pH 7.5 (measured in water). The pH of these soils is not much affected by organic acids and the reaction is dominated by the basic cations. This high pH is not influenced by location in the country, but lower pH occurs on rhyolitic tephras, especially in the south in high rainfall areas. Research shows that the pH drops rapidly after vegetation is established on the desert surfaces, such at the Geitasandur LandAid research site. There, the pH dropped from 0.3 to 0.5 units in 8 years after reclamation was initiated (Arnalds et al. 2013), and older restoration efforts show similar lowering trend.

The AUI database does not reveal a clear trend in pH with depth except that pH increases with depth in more pedons than it decreases with depth. The soil reaction is clearly related to the organic content of the soil horizons as is illustrated in Fig. 8.2. The regression equation with R^2 of 0.49 is highly significant, even though data are used from various depths of pedons scattered around the country. The variability, reflected by the scattering of the data points, can be related to the interacting effects of aeolian depositions and carbon accumulation, with lows at wetland sites far from aeolian deposition. At such locations, less amount of basic cations are released during chemical weathering and conditions are optimal for carbon accumulation. There are a cluster of data

O. Arnalds, *The Soils of Iceland*,
World Soils Book Series, DOI 10.1007/978-94-017-9621-7_8
© Springer Science+Business Media Dordrecht 2015

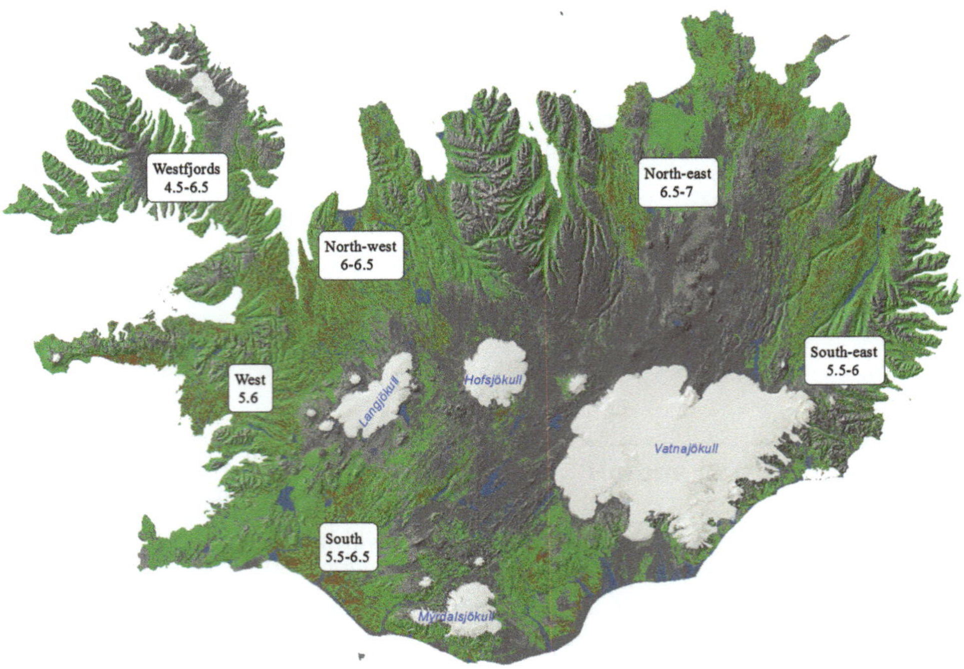

Fig. 8.1 A range for pH for surface horizons of Brown Andosols in the major regions of Iceland. Data from the AUI Database, prepared by SHB/OA; AUI

points at low carbon contents that both represent Vitrisols of high pH and also surface horizons with rhyolitic tephra in high rainfall areas showing lower pH.

Soil reaction measured in weak KCl solution (1 M) is generally 0.5–1.5 unit lower than pH measured in water. Soil reaction measured in KCl releases reserve acidity (Al^{3+} and H^+) from exchange sites, which is considerable according to these data. However, this reserve acidity is strongly related to the pH of the soil, and is most visible at higher pH, with a highly significant negative correlation with pH measured in water (Fig. 8.3).

An important property of Andosols is a strong response of pH to a weak NaF (1 M) solution, which often is used to identify the presence of andic soil materials, such as allophane and organo-mineral complexes (metal–humus complexes). This occurs because the active F^- ion replaces OH^- from the surfaces of the andic materials, thus increasing the pH. The pH (NaF) of the soils is predominantly <9.5 and is commonly between 10 and 11.5 (Fig. 8.4, see also table at the end of the chapter). The lower pH (NaF) values seen in the graph are both from Vitrisols and some Histosols with high carbon values but low $(Al + ½Fe)_{ox}$ indicating less andic influences.

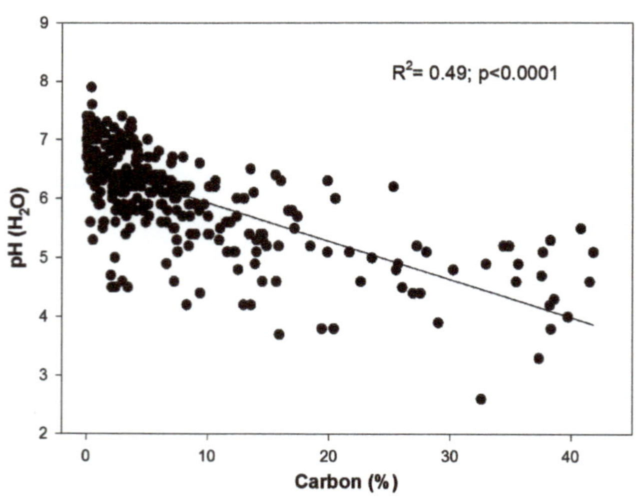

Fig. 8.2 The relationship between soil pH (H_2O) and carbon content for 319 soil horizons. Based on the AUI Soil Database

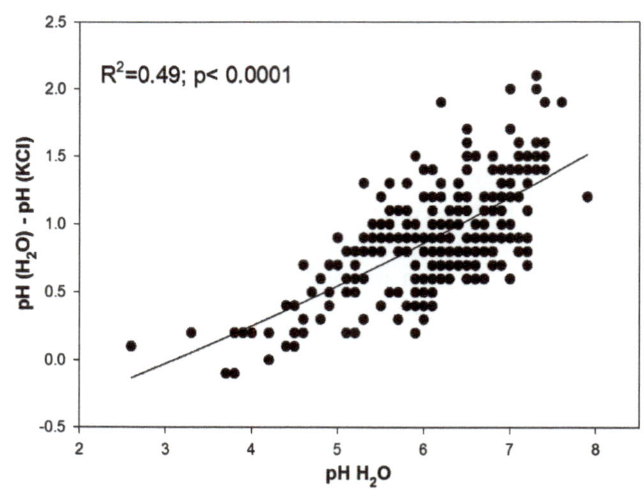

Fig. 8.3 The reserve acidity (pH H_2O–pH KCl) as a function of soil pH in H_2O, showing an increase with increased pH. Based on AUI Soil Database

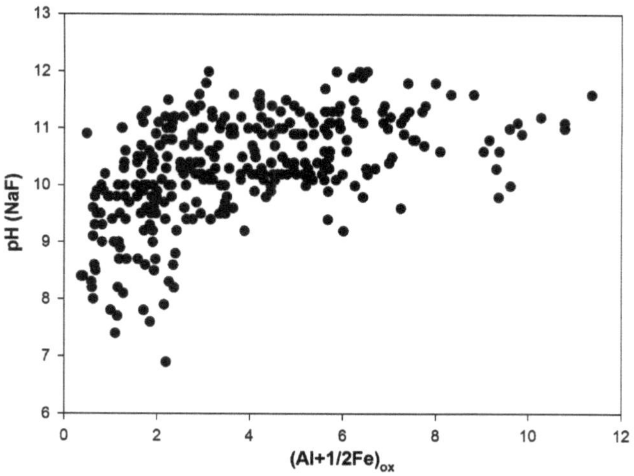

Fig. 8.4 pH (NaF) as a function of andic materials measured as $(Al + 1/2Fe)_{ox}$ (oxalate extractable aluminum and iron), showing a marked increase with andic soil properties. Most of the pH (NaF) is >9 and much of the values >10. Based on the AUI Soil Database

8.2 Charge Characteristics

Cation exchange capacity (CEC) is a measure of the capacity of the soil to retain cations in the soils which can be subsequently supplied to plant roots and organisms in the soil. CEC is therefore an important measure of soil fertility. It can either be determined directly or the exchangeable cations making up the CEC can be measured independently and added together. The cations Ca^{++}, Mg^{++}, Na^{+}, and K^{+} are the main "basic cations," while H^{+} and Al^{3+} are the "acid cations," contributing to soil acidity. The proportion of the sum of bases (SB) in the total CEC is termed "base saturation" (given as %).

There is a considerable difference in charge measured either as CEC or SB between the major soil classes in Iceland. Factors that affect the CEC include organic carbon, the amount of clay in the soils, and the pH. The AUI soil database contains mostly data for exchangeable bases (Ca^{++}, Mg^{++}, Na^{+}, and K^{+}) which are determined by using ammonium acetate buffered at pH 7. An average of 238 horizons from the AUI soil database gave an average SB of 24 $cmol_c\ kg^{-1}$ ($meq\ 100\ g^{-1}$) for Andosols and Histosols (not counting Vitrisols). The SB is generally highest for the organic horizons, often with 15–30 $cmol_c\ kg^{-1}$, but some histic horizons have SB > 50 $cmol_c\ kg^{-1}$. Surface horizons of Brown and Gleyic Andosols generally have SB of 10–20 $cmol_c\ kg^{-1}$ but the sum drops below 10 in areas receiving large amount of aeolian or tephra materials (low in clay and C). The Vitrisols (see Sect. 6.5) have SB 2–10 $cmol_c\ kg^{-1}$, which originates from significant, yet low, amounts of allophane and organic matter. A multiple regression relating SB to organic content, clay, and pH (H_2O) yielded a regression coefficient of 0.28 with $p < 0.0001$ ($n = 238$). This analysis (using the AUI database) was made for 238 horizons at various depths from evenly distributed pedon locations of Andosols and Histosols (excluding Vitrisols). This low coefficient, although highly significant, shows that there are many other factors that affect the charge characteristics than the amount of organic materials or clay, such as various humification stages of the organic matter (noted by Madeira et al. 2007b), but also the nature of the parent materials and possible presence of calcite in some horizons. Similarly made regression coefficients made for a database for European Andosols are generally higher (Madeira et al. 2007b).

The dominant exchangeable cation is calcium, reflecting the chemistry of the basalt glass in the soils. Calcium was generally 50–70 % of the exchangeable cations, with Mg^{++} also being abundant, often about $^{1}/_{3}$–½ of the exchangeable Ca^{++}. Exchangeable K^{+} is generally 0.5–2 $cmol_c\ kg^{-1}$ and Na^{+} 2–5 $cmol_c\ kg^{-1}$ according to the AUI database. Icelandic Andosols had the highest proportion of Ca^{++} in the exchangeable bases of the European Andosols studied under the COST-622 program (Madeira et al. 2007b), reflecting Ca rich volcanic parent material of soil in Iceland. KCl extractable Al^{3+} is generally low, indicating high base saturation of the soils as shown in the COST-622 study by Madeira et al. (2007b). A study by Arnalds et al. (1995) gives similar CEC values as the range in SB stated above.

The charge of Icelandic Andosols is characterized by the pH dependency of the charge. With rising pH, there is a proportional increase in the number of negative charges provided by soil colloids. CEC is therefore quite dependent on the pH during the extraction. It should therefore be stressed that the AUI method for determining the SB tends to overestimate the sum of exchangeable bases for the lower pH soils because of pH-dependent charge, but it still gives a good indication of the overall exchange properties. An analysis of three Icelandic pedons under the COST-622 program gives a good indication of the differences in pH-dependent charge properties (Table 8.1). Data for the uppermost 30–50 cm of soils are shown in the table. The pH and % C is typical for many soils in Iceland. The SB (measured at pH 7) ranges between 18 and 43 $cmol_c\ kg^{-1}$, and there is very little extractable Al^{3+}, resulting in the SB being the same as CEC_7, measured at pH 7. However, CEC measured at soil pH (CEC_S) is considerably lower for the EUR-N7 pedon due to the pH dependent charge. The EUR N8 and EUR N9 have about the same CEC_7 and CEC_S but the EUR N8 soil has higher pH and lower organic C, while the low difference for EUR-N9 (19.8 vs. 18) in spite of lower pH and high pH may in part be attributed to limited humification of the organic matter and relatively low allophane content.

It can be concluded from the paragraphs above that CEC of Icelandic soils is relatively favorable in most cases. This influences the fate of radioactive fallout materials such

Table 8.1 Exchange properties of upper horizons of the COST-622 Icelandic soil pedons

Horizon	Depth (cm)	pH (H_2O)	C (%)	Sum of bases	Al (KCl)	CEC_7	CEC_S	AEC
				cmol$_c$ kg^{-1} (meq 100 g^{-1})				
EUR N7 Ós, NW Iceland (Histic Andosol)								
O	0–5	6.3	19.9	42.6	0.14	42.7	17.8	6.4
Ah1	5–17	5.8	16.6	43.3	0.14	43.4	18.5	5.2
2Ah2	17–35/50	5.9	13.0	18.5	0.17	18.6	10.8	4.5
EUR N8 Auðkúluheiði, highland NW Iceland (Brown Andosol)								
Ah1	3–11/19	6.1	6.6	18.2	0.26	18.5	20.9	3.9
Ah2	11/19–21/27	6.5	5.7	18.7	0.13	18.9	20.9	4.2
Bw1	21/27–26/34	6.8	4.2	17.7	0.02	17.7	18.6	4.2
EUR N9 Hella, South Iceland (Gleyic Andosol)								
A1	0–55	5.7	10.0	19.4	0.4	19.8	18.0	5.1

Sum of bases measured with ammonium acetate at pH 7, CEC_7 is sum of bases plus extractable Al (KCl), CEC_S is measured at soil pH (compulsive exchange procedure). AEC is anion exchange capacity. Data compiled from Buurman et al. (2007). See Buurman et al. (2007) and Madeira et al. (2007b) for methods

as ^{137}Cs. Research shows that the amount of fallout cesium from the nuclear testing period is mostly dependent on precipitation, and that the majority of the ^{137}Cs (83 % on average) is retained in the uppermost 5 cm of the soil (CEC sites) (Sigurgeirsson et al. 2005). This is more than 40 years after the major fallout maximum in 1965. This shows the ability of Icelandic Andosols to retain polluting anions in surface horizons. Interestingly, Sigurgeirsson et al. (2005) found that the Vitrisols (vitric soils of the deserts) with only 2–5 % clay content and limited organic content retained 76 % of the ^{137}Cs fallout from the nuclear testing period, showing that the Vitrisols are unique for such recent sandy sediments. However, rapid aeolian deposition has the tendency to bury former surface horizons retaining fallout ^{137}Cesium.

Soils in Iceland also have anion exchange properties at low pH, primarily the Histic Andosols and the Histosols. Anion capacity can help retaining important anions such as NO_3^-. Madeira et al. (2007b) showed a general anion exchange capacity range for Icelandic pedons of 2–5 cmol$_c$ kg^{-1} (examples in the last column of Table 8.1). The pedons investigated by Madeira et al. (2007b) all had pH > 5.5 and presumably the anion exchange is considerably higher for the low pH soils.

8.3 Phosphorus Retention

Pronounced phosphorus retention is among the diagnostic properties of Andosols (Dahlgren et al. 2004) as is discussed in Chap. 5 on Andosols in general. Madeira et al. (2007a) showed that the P-retention in European Andosols was related to Al–humus complexes (organo-mineral complexes), allophane and ferrihydrite, but not to the total carbon content. Icelandic soils show typical tendencies for phosphorus retention, with general values >90 % for reasonably developed Brown and Gleyic Andosols with lower values for less weathered coarse-grained and silicic tephra horizons (20–50 %) (Arnalds et al. 1995). The COST-622 database (Buurman et al. 2007) indicates somewhat lower P-retention (80–90 %) for histic soil horizons compared to the Andosol horizons lower in organic matter (see table for nine pedons at the end of the chapter).

The Vitrisols of the desert surfaces (Vitricryands according to Soil Taxonomy) have considerably lower P-retention due to lower allophane and organo-mineral content, commonly in the range of 30–80 % (Arnalds and Kimble 2001), depending on composition. It is noteworthy, however, that soils of unweathered glaciofluvial sands fields have P-retention >20 %, but these materials do contain appreciable amounts of oxalate extractable Al and Fe, making the soils Andosols/Andisols (WRB/ST) in spite of the limited pedogenesis.

Helgason (2002) published a review paper on phosphorous in soils in Iceland. He showed that the range for organic P was from minimal to >2 g kg^{-1}. Helgason found that more of the total P (inorganic + organic) was bound organically outside of the active volcanic belt (48–53 %) than inside (17 % in South Iceland). The C:P ratio ranged from 66 to 399 with organic P closely related to carbon content, yet the bioavailability of P was only somewhat related to the carbon stock in the soil (Helgason 2002) as would be expected for Andosols.

The phosphorus fixing capacity of Andosols makes it necessary to use phosphorus fertilizers when these soils are cultivated. It is common practice to use 20–30 kg P ha^{-1} year^{-1} on Icelandic hay fields (e.g., 400 kg of 23-23-12 N—P_2O_5–K_2O fertilizer). The P levels build up in the soils, yet the continuous P fertilization seems necessary (Gudmundsson et al. 2005). Furthermore, fertilizers used for restoration efforts of severely degraded areas contain substantial proportion of phosphorus. However, there seem to be ample phosphorus turnaround in vigorous natural ecosystems.

8.4 Oxalate and Pyrophosphate Extraction

Acid oxalate and pyrophosphate extractions are widely used to account for short-range order clay minerals (allophane, imogolite, ferrihydrite) and to characterize the metal–humus complexes in Andosols. Iron and aluminum are most commonly determined in both the extracts, but silica mainly in the oxalate extract (Fe_{ox}, Al_{ox}, Si_{ox}, Al_{pyr}, Fe_{pyr}). These are very important chemical parameters for Andosols and give valuable information about the character of the soils (see Chap. 5). The oxalate extraction is a basis for the classification of Andosols. Oxalate dissolves both components (clay and metal–humus complexes), while pyrophosphate dissolves the humus complexes. Results for oxalate extractions are presented for nine pedons at the end of the chapter. Discussion of the results in relation to mineralogy is provided in the next chapter.

The aluminum extracted by oxalate (Al_{ox}) is in moderate amounts, commonly in the range of 1–3 % in Andosols (soils with vegetation cover), but higher (2–5.7) in the Breiðdalur Pedon (Southeast, Pedon 7 in Table 8.4 at the end of the chapter) and in some horizons of other pedons. Values for Vitrisols (soils of the deserts, Vitricryands according to ST) are generally about 1 % or lower (Arnalds and Kimble 2001). The Al extracted is both in allophane and bound with humus complexes. Pyrophoshate Al (Al_{pyr}), a measure of Al bound by metal humus complexes, has direct relationship with the total organic carbon, with higher Al_{pyr} values in organic rich horizons. Table 8.2 shows results of oxalate and pyrophosphate extractions from two COST-EU pedons, one relatively organic Histic Andosol and the other highland Brown Andosol (slow weathering) (Buurman et al. 2007; Garcia-Rodeja et al. 2007). These pedons are also depicted in the table at the end of Chap. 7 (descriptions) and this chapter. The Al_{pyr}/Al_{ox} ratio is relatively high for the Histic Andosol of Pedon 1, with ratios generally 0.3–0.4 and Fe_{pyr}/Fe_{ox} ratios even higher (0.5 to >1). This indicates that a considerable amount of the aluminum and iron is associated with metal–humus complexes in Pedon 1 (Ós), and that a considerable fraction of the carbon is in the form of aluminum–and iron–humus complexes. This pedon has relatively low pH, which facilitates higher Al^{3+} concentration in the soil solution needed for the metal–humus formation.

The Al_{pyr}/Al_{ox} and Fe_{pyr}/Fe_{ox} ratios are lower for Pedon 2 (Auðkúluheiði), which has lower organic contents, cooler climate and also higher pH (see Table 8.4 at the end of the chapter). This indicates that lower proportion of the Al and Fe is associated with metal–humus complexes. Cooler climate leads to slower decomposition of organic matter in Pedon 2,

Table 8.2 Oxalate and pyrophosphate extractions for two COST-622 pedons

Horizon	Depth (cm)	C	Al_{ox}	Fe_{ox}	$(Al + ½Fe)_{ox}$	Al_{pyr}	Fe_{pyr}	Al_{pyr}/Al_{ox}	Fe_{pyr}/Fe_{ox}
		%							
Pedon 1. Ós, Northwest. Histic Andosol. COST 622 EUR 07									
O	0–5	19.9	1.1	1.3	1.7	0.39	0.77	0.36	0.61
Ah	5–17	16.6	1.4	1.4	2.2	0.54	0.83	0.38	0.58
AC	17–35/50	13.0	2.1	6.1	5.2	0.83	1.17	0.39	0.19
2BC	35/50–65	17.4	2.4	0.6	2.7	1.16	0.78	0.49	1.24
3BC	65–73	6.7	2.9	0.6	3.2	0.83	0.44	0.29	0.69
3CB T	73–82	11.8	2.6	1.1	3.1	1.05	0.56	0.41	0.50
4Bw	82–90/100	8.5	5.2	0.8	5.6	1.08	0.75	0.21	0.96
4B/Cg	>90/100	0.5	0.3	0.6	0.6	0.02	0.15	0.06	0.25
Pedon 2. Auðkúluheiði, Northwest highlands. Brown Andosol. COST 622 EUR 08									
Ah1	3–11/19	6.6	1.9	1.4	2.6	0.40	0.48	0.21	0.34
Ah2	11/19–21/27	5.7	2.0	1.3	2.6	0.34	0.44	0.17	0.33
Bw1	21/27–26/34	4.2	2.7	1.5	3.5	0.29	0.30	0.11	0.19
2Bw2 T	26/34–37/42	2.0	1.5	0.7	1.9	0.13	0.12	0.08	0.16
3Bw3/4C	37/42–59/62	2.8	2.9	1.9	3.9	0.20	0.20	0.07	0.11
4C	>59/62	0.3	0.8	0.8	1.2	0.03	0.03	0.04	0.04

Carbon content shown for convenience. Data from University of Santiago de Compostela, Spain (see Buurman and van Doesburg 2007; Garcia-Rodeja et al. 2007). T is a horizon made of single tephra layer compared to Pedon 1, which has some enhancing effect on carbon buildup at Auðkúluheiði. Allophane clay content is low in Pedon 1 (lower pH slows/inhibits allophane formation, see next chapter), so organic materials are not much bound by allophane compared to metal–humus complexes.

The Fe_{pyr} values are very high compared to what is usual in Andosols in other countries. This can be related to the high content of the iron in the Icelandic parent materials (6–12 %; Jakobsson 2008), and high values are commonly found in soils where periodic reduction occurs (Gleyic/Histic Andosols), and where the iron is carried to areas of better aeration, e.g., at the boundary of coarse tephra layers.

The $(Al + ½Fe)_{ox}$ is generally above the 2 % limit (Table 8.4) for the Brown Andosols and Gleyic Andosols used for Andosol classification in WRB and Soil Taxonomy. The glass counts for these soils are generally high (Stoops and Gerrard 2007; Stoops et al. 2008), which would lower the $(Al + ½Fe)_{ox}$ limit under WRB and Soil Taxonomy below 2 %. The Histic Andosols and Histosols have lower $(Al + ½Fe)_{ox}$ but still generally >2 %, especially the Histic Andosol. The >20 % C criterion overrides the oxalate criterion in the classification, defining the Histosols, which still have many chemical properties typical of the Andosols. The $(Al + ½Fe)_{ox}$ is lower for the Vitrisols, yet often relatively high (>1 %), and these soils are mostly classified as Andosols under the WRB and Soil Taxonomy as is discussed in Chap. 6 on classification.

8.5 Carbon and Nitrogen

8.5.1 General Carbon Levels

Organic carbon accumulation is one of the major characteristics of Andosols in the world (Nanzyo et al. 1993). This is quite evident in Iceland, of course, but the unusually active geomorphic environment, with continuous aeolian deposition, periodic tephra deposition, and the presence of extensive desert surfaces, makes the carbon content much more variable than otherwise would be expected. The allophanic Andosols of the world often stabilize at about 6 % C, but alu-andic (metal–humus complexes) soils often have more organic matter. In Iceland, there is a full range from nearly zero C in Vitrisols (which classify as Andosols according to WRB and Soil Taxonomy) to >20 % C in organic soils with andic soil properties. This carbon accumulation tendency is part of the andic soil properties, and therefore the division between Andosols and Histosols is set at 25 % C according to the WRB (and also Soil Taxonomy), but the Icelandic system uses the 20 % limit as a boundary.

The carbon content of the Brown Andosols (dryland soils under vegetation) ranges generally between 2 and 5 %, but is higher in surface horizons in fertile ecosystems such as the birch forests and grasslands. The vegetation cover is an important factor influencing the carbon accumulation in addition to soil functions. But the overriding factors in Iceland are aeolian deposition, drainage, and long-term land use (grazing). The cold climate of Iceland is an additional factor contributing to the accumulation of organic matter as decomposition tends to be slow. Yet, organic accumulation will only occur with good vegetation cover. The soils of heavily grazed heathlands are generally limited by limited organic input as aboveground productivity is low and/or removed by grazing. This effect is, however, amplified by the influence of aeolian inputs, which lowers the organic content of each horizon. Carbon levels (in %) within each class of soil are therefore lower in areas receiving large amounts of aeolian inputs. Brown Andosols and Gleyic Andosols often have as low as 2–3 % C in surface horizons close to active aeolian sources, while the levels exceed 10 % far from the aeolian sources in fertile systems.

The Histosols have >20 % C in surface horizons by definition. They are found at poorly drained landscape positions where aeolian deposition is limited (see Chap. 6). The carbon levels in the Vitrisols of barren surfaces is quite different from the Histosols and Andosols as they are soils of limited vegetation cover and therefore have limited organic input to the soils.

The depth distribution of the carbon can be described as erratic. The environmental conditions vary considerably during each time segment of soil formation as the soil is gradually becoming deeper. This leads to surprisingly different carbon levels between soil horizons—low carbon content when aeolian deposition is rapid and especially low in thick tephra layers, but higher when deposition is slow (calm weathering environment). This is illustrated in Fig. 8.5 where carbon contents of several soil horizons have been plotted.

Oskarsson et al. (2004) used the AUI database to see the average depth distribution of carbon for the different soil types. The results for Sandy and Cambic Vitrisols, Brown Andosols, and Gleyic Andosols are shown in Fig. 8.6, which shows the pronounced difference between the Vitrisol and Andosol classes. Another noteworthy point is that there is a marked decrease in organic carbon at around 20 cm depth on average. This decrease has been explained by increased aeolian activity during the Middle Ages, when severe erosion of existing soils (mostly Brown Andosols) caused more rapid aeolian redistribution and deposition. The vertical distribution curve for the Histic Andosols and Histosols follows a similar trend (not shown on the figure).

8.5.2 Carbon Stocks—Accumulation

One can consider the surface carbon stocks as the energy reserves of the ecosystems. These reserves are not only important for the immediate ecosystem drawing from these

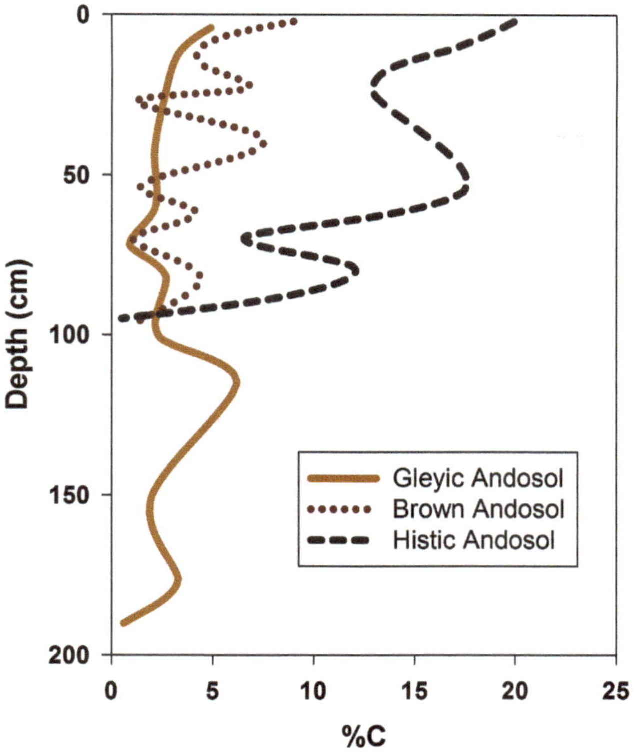

Fig. 8.5 Organic carbon distribution for three selected soil pedons, a Gleyic Andosol (S Iceland), Brown Andosol (NE Iceland) and Histic Andosol (NW Iceland). The Gleyic and Brown Andosols have tephra layers low in organic content. Data from the AUI Soil Database

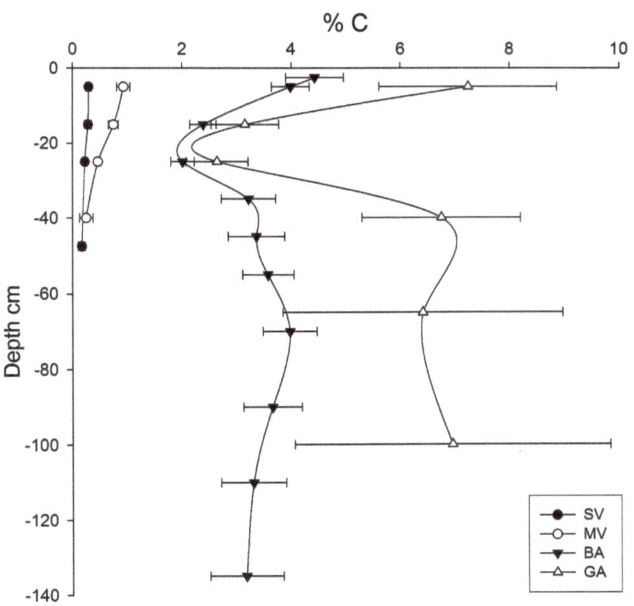

Fig. 8.6 Average vertical distribution ofcarbon for Sandy and Cambic Vitrisols and Brown and Gleyic Andosols. There is a clear decrease in carbon content of the Andosols that coincides with aeolian redistribution of soils during the Middle Ages. Data from the AUI database, based on Oskarsson et al. (2004)

reserves; the fate of the soil carbon affects the global carbon cycle. The cycle is, however, not only affected through the organic carbon, but also in relation to chemical weathering, which can adsorb large quantities of CO_2, especially when the rocks are rich in Ca, which combines with CO_2 promoting the formation of $CaCO_3$ in rocks and oceans. Gislason published a book (2012) on the carbon cycle with special reference to Icelandic conditions and especially the geochemical aspect of the cycle (see also Gislason 2008; Gislason et al. 2009), but chemical weathering is discussed in more detail in the next chapter of this book.

Soils store more carbon on a global scale than are found in the atmosphere and aboveground vegetation combined—soils are extremely important component in the carbon cycle on Earth. Release of carbon from soils increases greenhouse gas levels in the atmosphere, but a substantial proportion of annual greenhouse gas emissions could potentially be absorbed to terrestrial ecosystems, both vegetation and soils (Lal 2004). Severe depletion of these reserves has occurred in Iceland where Vitrisols have taken the place of Andosols due to erosion, and many of the ecosystems, such as the poor heathlands, have reduced carbon stocks in surface horizons.

Andosols contain the largest amount of carbon of all dryland soils, often >30 kg C m^{-2} (Batjes 1996; Eswaran et al. 1993). Icelandic Andosols contain the characteristic high amounts of carbon, often ranging between 15 and 80 kg C m^{-2}. Oskarsson et al. (2004) estimated the total carbon stocks of Iceland to be 2.1×10^9 tons and that $120–500 \times 10^6$ tons C have been lost due to land degradation since the Settlement. However, the aeolian additions cause the surface to rise, continuously accumulating and burying carbon from the atmosphere. The rates vary between ecosystems and rates of aeolian deposition, commonly ranging from 0.005 to 0.03 kg C m^{-2} $year^{-1}$ (based on Nytjaland Database, AUI Soil Database, and the AUI IGLUD Database; Jon Gudmundsson and O. Arnalds AUI unpublished data). Gisladottir et al. (2010) reported long-term carbon accumulation of the same order on heavily used land (grazed), since the twelfth century in Southwest Iceland (0.017–0.030 kg C m^{-2} $year^{-1}$), resulting from the continuous aeolian burial. Ritter (2007) reported rates up to 0.023 kg C m^{-2} $year^{-1}$ in relation of tree plantation in heath soils in Iceland. These rates are higher than commonly reported for the world's Andosols, but Zehetner (2010) suggested an average of 0.01 kg C m^{-2} $year^{-1}$ as average accumulation for the world's volcanic soils based review of published research. Jon Gudmundsson and O. Arnalds estimated carbon accumulation in Icelandic soils of the order 200,000–900,000 tons C $year^{-1}$ in dryland soils (AUI unpublished data), which is a substantial proportion of the annual release from fossil fuel burning in Iceland. In addition, there is considerable accumulation in vegetation, especially in relation to restoration of the birch forests and

Fig. 8.7 The Geitasandur LandAid restoration research area consists of 40 one ha plots with 10 treatments replicated four times, including a control. Land treated with grass seeds and fertilizer to the *left*, untreated land on the *right*. The research shows carbon accumulation of 0.04–0.08 kg C m^{-2} $year^{-1}$ during the first years, a rate that can potentially be maintained >200 years (see Arnalds et al. 2013)

other tree plantations (Bjarnadottir et al. 2007, 2009; Snorrason et al. 2002). However, heavy land use and partially open land (erosion spots, etc.) can cause considerable and measurable release of CO_2 from these systems as well (Einarsson 2013; Gudmundsson and Oskarsson 2014).

As much of Iceland is poorly vegetated, there are many restoration efforts being carried out (see the final chapter of the book). The nature of the soils and cold climate leads to rapid organic accumulation in the soils after restoration efforts are initiated (Arnalds et al. 2000, 2013; Aradottir et al. 2000) (Fig. 8.7). Common rates found are 0.04–0.08 kg C m^{-2} $year^{-1}$ or 400–800 kg ha^{-1} $year^{-1}$, which can be maintained over a long time, potentially >200 years. These rates are among the highest found for simple management practices in the world literature—a result of having deserts under humid cold climate with andic soil properties which are factors that all favor carbon accumulation and carbon rich systems when fully restored. Currently, restoration efforts add carbon accumulation in soils in Iceland of the order of 50,000 tons C annually (Hallsdottir et al. 2012).

8.5.3 The Icelandic Wetlands in Relation to Carbon Budgets

The wetlands, ranging from organic Histosols to Gleyic Andosols with low organic content near active aeolian sources and volcanoes, are important ecosystems on a national scale due to their rich biodiversity, water regulation, and ecosystem services to other systems (Fig. 8.8). They are important international areas for the preservation of many bird populations.

Wetland soils are considered to be 5,800–8,000 km^2, depending on how they are accounted for, with nearly half of this area below 200 m elevation (see Chap. 4). However, they are also very important for agriculture, mainly for making hay as winter fodder. They have been extensively drained for this purpose (Oskarsson 1998; Thorhallsdottir et al. 1998), but about 45 % of the hay making land ('tún') are drained wetlands (Helgadottir et al. 2013). Gudmundsson et al. (2013) estimated that 3,400–3,900 km^2 of the wetlands have been drained. These are drained with about 31,600 km long ditches and additional 61,600 km of subsurface channeling (AUI unpublished data; see further discussion on wetland disturbance in Chap. 12.)

The AUI operates a Wetland Center. AUI has made estimates of CO_2, CH_4, and N_2O emissions from the drained wetlands. The values are rather staggering. Gudmundsson and Oskarsson (2014) measured emissions of 4–8.2 tons C ha^{-1} $year^{-1}$ (14.6–30.3 tons CO_2 ha^{-1} $year^{-1}$; see also Maljanen et al. 2010). These measured emissions are of the same order as measured in the other Nordic countries (Maljanen et al. 2010). Applying a mean value to the drained wetlands of 3,440 km^2 results in release of 1.4–4.8 million tons of C each year to the atmosphere (5–10.5 million tons CO_2) (see Gudmundsson and Oskarsson 2014). These values are considerably larger than the 4.4 million tons of CO_2 released by other sectors (industry, transportation, etc.), casting doubt on the justification for the extensive drainage of Icelandic wetlands.

Draining peatlands outside of Iceland often leads to oxidation, or burning of the peat, when oxygen gets access to the high organic reserves. Consequently, the surface can start to subside, even from several meters to few centimeters, with time. Interestingly, the oxidation of the drained

Fig. 8.8 Typical wetland area, in part drained for agricultural production. Hay is being made on the fields. Extensive afforestation areas in background with introduced species. Relatively undisturbed wetland in the foreground. *Photo* © Tomas Gretar Gunnarsson, University of Iceland

wetlands in Iceland has not caused much subsidence or lowering of the surface, which is attributed to the presence of the volcanic materials in the matrix (Bartoli and Burtin 2007). No subsidence has been reported in the literature to date, but AUI staff has made some vague observations of subsidence at experimental plots.

Several successful attempts have been made to restore wetlands (Fig. 8.9). Groups and institutes engaged in such activities include the road authorities (to counteract wetland disturbance from road construction), a large aluminum smelter company (to balance carbon emissions), farmers and other landowners and NGOs (see Aradottir and Hagen 2013; Oskarsson 2011). However, these efforts are small to date, compared with the extensive drainage of the wetlands.

8.5.4 Nitrogen

Nitrogen is often the most limiting nutrient in agriculture and in early stages of ecosystem development following land degradation in Iceland. Fertilizers are used extensively for haymaking in Iceland. Common fertilizer applications are of the order of 80–140 kg N ha^{-1}, a total of about 15,000 tons on about 1,250 km^2 hay fields (average of 120 kg N ha^{-1}; data from the Farmers Association; www.bondi.is). These fertilizer rates are similar to those found in other Nordic countries, however, lower than is commonly found in Denmark. Nitrogen pollution from fertilizers has not been documented to any extent and is believed to be minimal due to the low density of the agriculture and relatively moderate applications levels. Fertilizer applications are based on extensive fertilizer experiments which have been published in Icelandic (mainly the annual Agricultural Congress in Iceland, 'Fræðaþing'), but also in some international journals (see e.g., Gudmundsson et al. 2004, 2005). The methods are in need for revisions and increased scrutiny (Gudmundsson and Sveinsson 2011). There has been considerable research and modeling of nitrogen mineralization rates in the soils (e.g., Palmason et al. 1996) and in relation to afforestation, revegetation and land restoration research (e.g., Oskarsson et al. 2006; Oskarsson and Brynleifsdottir 2009; Thorvaldsson et al. 2009).

The AUI IGLUD Database and soil databases provide a good opportunity to explore total nitrogen on regional scales and in relation to factors such as aeolian additions to the soils (Table 8.3). The nitrogen stocks in the top 30 cm of Brown Andosols under heathland (see Chap. 4 on vegetation classes) are from 4,500 to almost 11,000 kg N ha^{-1}. Rich heathland has higher N on average than the poor heathland except where aeolian additions are rated very high, where both vegetation classes have relatively low nitrogen levels in the top 30 cm of soil. The pattern for N stocks in the wetlands are noteworthy, and are explained by the low bulk density for the more organic soils (low aeolian additions) and aeolian additions (top two rows). There is a general tendency of higher N with less aeolian additions per depth increment as would be expected, underlining the importance of the aeolian additions on the ecosystems.

Fig. 8.9 Restoration of a wetland at an AUI experimental farm in West Iceland. The ditches are filled in with soil materials. Restoration cuts down the severe greenhouse gas emissions and brings back the functions of the system in relatively short time (few years). The road authorities, an aluminum smelter company, private landowners, NGOs, and the government sector (including the Agricultural University of Iceland) have engaged in wetland restoration in Iceland. *Photo* © Hlynur Oskarsson, AUI

Table 8.3 Nitrogen stocks in the top 30 cm of poor heathlands, rich heathlands, and wetland vegetation categories (see Chap. 4 on vegetation types)

Aeolian deposition category	Poor heathland		Rich heathland		Wetland	
	BD[a] (g cm^{-3})	Stock (kg N ha^{-1})	BD[a] (g cm^{-3})	Stock (kg N ha^{-1})	BD[a] (g cm^{-3})	Stock (kg N ha^{-1})
Very high	0.74	4,516	0.74	4,643	0.63	8,145
High	0.66	9,551	0.67	8,746	0.58	11,802
Medium	0.69	7,788	0.62	11,253	0.53	12,749
Low	0.60	10,704	0.53	14,347	0.13	4,839

Poor and rich heathlands have Brown Andosols, but the wetlands a range from Histosols (>20 % C, Histic Andosols (12–20 % C) to Gleyic Andosols (<20 %C). Based on the Agricultural University of Iceland IGLUD and Soil Databases

[a] Calculated BD using a regression equation based on AUI data: BD = 0.812 − (C % × 0.0203)

8.6 Trace Elements

As the parent materials of soils in Iceland are mainly basaltic volcanic materials, a suite of elements can be expected to be released to the soil solution by weathering and as part of various organo-mineral complexes. However, mobility of many ions are limited because of high CEC, but the concentration of ions that get mobilized under reduced conditions, such as Fe and Mn, may both increase and decrease, depending on the drainage conditions. Tanneberg and Jahn (2007) concluded that volcanic soils have in general high sorption capacity for heavy metals at different pH values, with strong sorption of Pb and Cr. Many research efforts have been made to monitor trace elements in fodder-hay and in domestic animals (Johannesson et al. 2004a, b, 2005, 2007; Hardarson et al. 2006; Thorvaldsson and Gudmundsson 2006). Thorvaldsson and Gudmundsson (2006) concluded that microelements are not limiting in grass production in Iceland, but copper (Cu), manganese (Mn), and molybdenum (Mo) need to be monitored. Iron levels tend to be high in hay fodder, generally 100–1,000 mg kg^{-1} (Johannesson et al. 2007), which is a direct result of aeolian deposition of basaltic glass high in iron (10 %). Gudmundsson and Thorsteinsson (1980) reported low values for Cu in hay. Cu has a tendency to be fixed by organo complexes, which are characteristic of many of the Icelandic Andosols. Ragnarsdottir and Hawkins (2006) noted both low levels of Cu in the soils and also unfavorable Mn/Cu ratio. However, Panek and Kepinska (2002), who investigated four sites in Iceland, reported rather high Cu levels compared to soils in northern and Central Europe, or 34–150 mg kg^{-1}. High Mn values are due to the basaltic chemistry and reduction in wetland soils.

Zink (Zn) is easily adsorbed to organo-mineral complexes and seems to be strongly adsorbed on the surface of iron hydroxides and iron metal humus complexes in Iceland (unpublished research, see Arnalds and Guicharnaud 2008). Lead levels (Pb) have been reported as rather low for soils in Iceland (Panek and Kepinska 2002).

The mobile form of molybdenum (Mo) is primarily Mo^{IV}, which forms stable bonds with Al- and Fe oxides and hydroxides. The Mo levels measured in hay in Iceland varies considerably and is sometimes very low (Johannesson et al. 2005, 2007), most likely due to the presence of ferrihydrite, allophane, and metal–humus complexes. Another trace element, selenium, has long been of concern in Icelandic agriculture, as the Se concentrations in Icelandic hay is low (e.g., Johannesson et al. 2005; Hardarson et al. 2006).

8.7 Biology

The biological activity is the driver of nutrient cycling in the soils; it affects water chemistry and weathering and is fundamental to the function of ecosystems. Soil ecology, which deals with microflora (fungi and bacteria), microfauna (e.g., protozoa and nematodes), mesofauna (e.g., collembolan and mites), and macrofauna (e.g., isopods and earthworms) is a young science in Iceland and some of these sectors have not been addressed in detail, especially the microfauna. There are several publications, mainly in Icelandic, on various aspects of the mesofauna and macrofauna of the soils, mainly Collembola and earthworms. These include lists of sampled of invertebrates (Gudleifsson and Bjarnadottir 2002) and the long-term influences of fertilizers on invertebrates (Gudleifsson 2002). The Collembola is the best studied species group. Recent study shows that Collembola species richness in general is lower in Iceland than under similar climatic conditions elsewhere, except in undisturbed systems such as old birch woodlands, where species richness is higher than in comparable birch forests in Norway (Fjellberg 2007). Fjellberg (2007) also found high species richness in frequently disturbed natural systems such as sea shores and river banks. The low species richness is attributed to degradation of the most common ecosystems in Iceland.

Sigurdardottir (1994), Gudleifsson (2007), and Gudleifsson and Olafsson (1981) reviewed the occurrence and importance of earthworms in soils of Iceland, contributing to such factors as nutrient cycling and physical structure of the soils. There have been 11 earthworm species identified, which is considerably fewer species than in the other Nordic countries (Sigurdardottir and Thorvaldsson 1994) with higher number of earthworms in fertile garden systems than natural sites (Bengtsson et al. 1975). Gudleifsson (2002) showed that fertilizers and lowering of the pH severely affected earthworm populations. Gudleifsson (2005) also has published papers on beetle species (*Coleoptera*) in Iceland.

There has been considerable research on soil fungi, ectomycorrhiza, and their effects of seedlings and survival of various plant species (e.g., Oddsdottir et al. 2010a, b, c), Soil ecology has also received considerable attention in relation to land reclamation and restoration (e.g., Oddsdottir et al. 2008). Oskarsson and coworkers (e.g., Enkhtuya et al. 2003; Oskarsson 2012) have studied the function of mycorrhiza in soils in Iceland. They are extremely important for the function of the lymegrass (*Elymus arenarius*), which is fundamental for sand stabilization in Iceland, but also in birch (Oskarsson 2010, 2012).

Guicharnaud et al. (2009, 2010) studied microbial activity in relation to temperature increase, seasonal changes, and fertilizer treatments. They studied factors such as respiration, nutrient availability, microbial biomass carbon, arylphosphatase, and dehydrogenase activity and showed that the soil temperature regime affected the soil microbial biomass carbon sensitivity to temperatures. When soils where sampled from the cryic temperature regime, a decreasing soil microbial biomass was detected when temperatures rose above the freezing point. Frigid soils, sampled from milder climatic conditions, where unaffected by difference in temperatures. Nitrogen mineralization did not change with temperature.

8.8 Chemical Data for Nine Selected Soil Pedons

Chemical data for nine selected soil pedons (a total of 66 horizons) is presented in the table below (Table 8.4). These are the same pedons that have soil and site descriptions in Chap. 7 with photographs, so the reader is referred to that chapter for visualization and more physical details about the pedons. The data is derived from various sources: European COST-622 profiles for Pedons 1–3 (see book edited by Arnalds et al. 2007); two from the Ph.D. work of the author (Pedons 4 and 5, Arnalds 1990; Arnalds et al. 1995), two profiles from AUI unpublished data (pedons 6 and 7) and desert profiles (pedons 8 and 9) published by Arnalds and Kimble (2001). Methods for soil analysis are comparable between these different studies, except for the CEC, which is measured under different pH conditions, which greatly affects the results. In some pedons (6 and 7), (SB) is given but not the CEC.

Table 8.4 Chemical properties of nine selected pedons

Horizon	Depth (cm)	C (%)	pH H_2O	pH NaF	P-ret (%)	CEC ($cmol_c\ kg^{-1}$)	Si_{ox}	Al_{ox}	Fe_{ox}	Al_{ox} + ½Fe_{ox}
									%	
Pedon 1. Ós, Northwest. Histic Andosol. COST 622 EUR 07										
O	0–5	19.9	6.3	8.8	75	59	0.4	1.1	1.3	1.7
Ah	5–17	16.6	5.8	9.4	70	58	0.5	1.4	1.4	2.2
AC	17–35/50	13.0	6.0	10.6	93	49	0.9	2.1	6.1	5.2
2BC	35/50–65	17.4	5.8	10.8	95	51	0.7	2.4	0.6	2.7
3BC	65–73	6.7	5.8	11.0	92	35	1.1	2.9	0.6	3.2
3CB T	73–82	11.8	5.5	10.8	94	50	1.0	2.6	1.1	3.1
4Bw	82–90/100	8.5	5.7	11.1	98	66	2.2	5.2	0.8	5.6
4B/Cg	>90/100	0.5	5.3	8.6	51	27	0.2	0.3	0.6	0.6
Pedon 2. Auðkúluheiði, Northwest highlands. Brown Andosol. COST 622 EUR 08										
O	0–3	–	–	–	–	–	–	–	–	–
Ah1	3–11/19	6.6	6.1	10.6	90	32	1.0	1.9	1.4	2.6
Ah2	11/19–21/27	5.7	6.4	10.5	92	29	1.1	2.0	1.3	2.6
Bw1	21/27–26/34	4.2	6.7	10.6	95	34	1.5	2.7	1.5	3.5
2Bw2 T	26/34–37/42	2.0	6.9	10.4	78	18	0.9	1.5	0.7	1.9
3Bw3/4C	37/42–59/62	2.8	6.9	10.3	96	32	2.0	2.9	1.9	3.9
4C	>59/62	0.3	7.1	9.7	48	11	0.5	0.8	0.8	1.2
Pedon 3. Hella, South Iceland. Gleyic Andosol. COST 622 EUR 09										
Ah	0–55	10.0	5.7	10.3	87	42	1.1	2.3	1.1	2.8
2C T	55–60	–	–	–	–	–	–	–	–	–
3H	60–95	19.9	5.1	9.8	40	51	0.6	1.6	0.6	1.9
3C	95–100	2.4	5.6	9.8	53	15	0.3	0.8	0.4	1.0
4H	100–230	13.0	4.2	10.2	96	48	0.7	2.1	0.7	2.4
Pedon 4. Goðafoss, Northeast. Brown Andosol (Arnalds 1990; Arnalds et al. 1995)										
A1	0–4	9.0	5.9	10.4	97	41	2.5	2.8	4.3	4.9
A2	4–12	5.0	6.5	10.2	95	32	2.5	2.5	4.2	4.6
A3	12–20	4.6	6.3	10.2	95	40	2.5	2.4	3.8	4.3
A4 (t1)	20–26	6.6	6.3	10.1	86	27	1.7	1.7	3.0	3.2
Bw1 T	26–29	1.2	6.6	10.0	–	9	1.1	1.0	1.4	1.7
Bw2 (t1)	29–41	7.1	6.7	9.9	99	44	2.3	2.2	4.5	4.4
Bw3	41–49	5.1	6.7	10.2	98	42	3.4	2.9	5.3	5.6
Bw4 T	49–57	1.4	6.7	10.6	66	11	1.1	1.1	0.6	1.4
Bw5	57–65	4.1	6.6	10.4	96	40	2.8	3.2	4.5	5.4
Bw6	65–70	2.1	6.6	10.1	98	25	2.0	2.2	3.2	3.8
Bw7 T	70–73	1.4	6.6	10.1	–	14	1.4	1.5	1.2	2.1
Bw8 (t1)	73–91	4.4	6.7	10.3	98	44	3.5	5.9	6.8	9.3
2Bw9	91–101	1.3	6.6	10.1	94	26	2.7	3.1	4.5	5.3
2C	101–121	0.1	6.7	9.8	47	14	0.9	1.0	1.8	1.9
Pedon 5. Þingvallasveit, Southwest. Brown Andosol (Arnalds 1990; Arnalds et al. 1995)										
A1	0–12	7.9	5.7	11.0	99	32	2.2	3.1	5.0	5.6
A2	12–28	7.6	6.1	10.7	99	45	2.7	4.5	6.5	7.8
Bw1	28–61	7.4	6.0	10.8	99	44	2.5	5.6	7.1	9.2
Bw2	61–68	4.2	6.0	11.1	98	32	2.7	5.5	3.5	7.2

(continued)

Table 8.4 (continued)

Horizon	Depth (cm)	C (%)	pH H_2O	pH NaF	P-ret (%)	CEC ($cmol_c$ kg^{-1})	Si_{ox}	Al_{ox}	Fe_{ox}	Al_{ox} + ½Fe_{ox}
									%	
2Bw3	68–87	2.1	6.1	10.5	99	25	2.8	3.4	2.6	4.7
2C	87–142	0.8	6.0	10.3	–	12	2.1	2.2	2.1	3.2
Pedon 6. Möðruvellir, North. Histic Andosol. AUI soil database										
A1 (H)	0–30	14.6	5.4	9.4	n.a.	18	1.5	1.8	7.8	5.6
2O1	30–55	12.5	4.8	9.8	n.a.	13	0.9	1.6	0.8	2.0
2O2	55–83	7.4	5.8	10.7	n.a.	15	1.1	1.7	0.6	2.0
2O3	83–98	10.0	5.4	10.4	n.a.	22	0.7	1.2	0.6	1.6
3C T	98–104	1.4	5.6	10.9	n.a.	2	0.4	0.4	0.1	0.5
4O1	104–145	35.4	4.6	8.6	n.a.	37	0.3	1.1	1.2	1.7
4O2	145–180	37.6	5.1	8.7	n.a.	69	0.3	1.3	1.4	2.0
4O3	180–200	34.8	5.2	9.0	n.a.	65	0.6	1.2	1.4	1.9
4O4	200–260	28.0	5.1	9.3	n.a.	58	0.5	1.5	0.8	1.8
4O5	260–300	27.2	5.2	10.0	n.a.	35	1.0	1.9	0.7	2.2
4O6	300–350+	41.8	5.1	7.4	n.a.	86	0.2	0.6	1.0	1.1
Pedon 7. Breiðdalur, Southeast. Gleyic Andosol. AUI soil database										
A1	0–11	9.3	5.8	11.2	n.a.	17	0.9	2.0	2.9	3.5
A2-t	11–26	7.2	6	11.6	n.a.	42	1.1	2.4	3.7	4.2
Bw1-t	26–37	8.2	6.1	11.9	n.a.	14	1.7	3.7	4.9	6.2
Bw2	37–47	7.2	6.2	11.9	n.a.	10	2.0	3.9	5.0	6.4
Bw3	47–70	10.1	6.2	12.0	n.a.	14	2.4	5.4	1.9	6.3
Bw4	70–82	10.6	6.2	12.0	n.a.	17	2.5	5.7	1.6	6.5
2Bw1	82–100	5.4	6.1	11.0	n.a.	9	2.3	5.1	1.4	5.9
2Bw2	120–120+	0.7	6.2	11.0	n.a.	23	0.3	0.9	0.7	1.2
Pedon 8. Sigalda, South highlands. Cambic Vitrisol (Arnalds and Kimble 2001)										
C	0–2	–	–	–	–	–	–	–	–	–
2A1	2–22	0.1	7.1	9.7	49	11	0.8	1.0	1.9	2.0
2A2	22–34	0.08	7.2	9.5	38	12	0.7	1.0	1.9	2.0
3C	34–50+	0.05	7.2	9.6	52	16	0.6	0.8	1.9	1.8
Pedon 9. Mýrdalssandur, South. Sandy Vitrisol (Arnalds and Kimble 2001)										
A	0–4	0.09	6.8	9.8	32	3	0.5	0.6	1.5	1.3
C1	4–8	0.05	6.9	9.8	35	3	0.4	0.4	1.1	0.9
C2	8–13	0.02	6.7	6.6	17	3	0.3	0.2	0.5	0.4

These pedons are described at the end of Chap. 7. The source of the data is explained in the text. T marks horizons made of single tephra layers, t distinct minor tephra layers within horizon

References

Regarding punctuation and Icelandic characters in citations: See note on punctuation in the Preface

Aradottir AL, Hagen D (2013) Ecological restoration: approaches and impacts on vegetation, soils and society. Adv Agron 120:173–222

Aradottir AL, Svavarsdottir K, Jonsson TH, Gudbergsson G (2000) Carbon accumulation in vegetation and soils by reclamation of degraded areas. Icel Agric Sci 13:99–113

Arnalds O (1990) Characterization and erosion of Andisols in Iceland. Unpublished PhD dissertation, Texas A&M University, College Station, Texas

Arnalds O, Kimble J (2001) Andisols of deserts in Iceland. Soil Sci Soc Am J 65:1778–1786

Arnalds O, Guicharnaud R (2008) Soils and public health. Agric Sci Congr (Fræðaþing landbúnaðarins) 2008:59–70 (in Icelandic)

Arnalds O, Hallmark CT, Wilding LP (1995) Andisols from four different regions of Iceland. Soil Sci Soc Am J 59:161–169

Arnalds O, Gudbergsson G, Gudmundsson J (2000) Carbon sequestration and reclamation of severely degraded soils in Iceland. Icel Agric Sci 13:87–97

Arnalds O, Bartoli F, Buurman P, Oskarsson H, Stoops G, Garcia-Rodeja E (eds) (2007) Soils of volcanic regions in Europe. Springer, Heidelberg

Arnalds O, Orradottir B, Aradottir A (2013) Carbon accumulation in Icelandic desert Andosols during early stages of restoration. Geoderma 193–194:172–179

Bartoli F, Burtin G (2007) Organo-mineral clay and physical properties in COST-622 European volcanic soils. In: Arnalds O, Bartoli F, Buurman P, Oskarsson H, Stoops G, Garcia-Rodeja E (eds) Soils of volcanic regions in Europe. Springer, Heidelberg, pp 469–491

Batjes NH (1996) Total carbon and nitrogen in the soils of the world. Eur J Soil Sci 47:151–163

Bengtsson SA, Nilsson A, Nordström S, Rundgren S (1975) Habitat selection of lumbricids in Iceland. Oikos 26:253–263

Bjarnadottir B, Sigurdsson BD, Lindroth A (2007) Estimate of annual carbon balance of a young Siberian larch (*Larix sibirica*) plantation in Iceland. Tellus B 59:891–899

Bjarnadottir B, Sigurdsson BD, Lindroth A (2009) A young afforestation area in Iceland was a moderate sink to CO2 only a decade after scarification and establishment. Biogeosciences 6:2895–2906

Buurman P, van Doesburg JDJ (2007) Laser-diffraction grain-size analyses of reference profiles. In: Arnalds O, Bartoli F, Buurman P, Oskarsson H, Stoops G, Garcia-Rodeja E (eds) Soils of volcanic regions in Europe. Springer, Heidelberg, pp 453–468

Buurman P, Bartoli F, Basile A, Füleky G, Garcia Rodeja E, Hernandez Moreno J, Madeira M (2007) The physico-chemical database. In: Arnalds O, Bartoli F, Buurman P, Oskarsson H, Stoops G, Garcia-Rodeja E (eds) Soils of volcanic regions in Europe. Springer, Heidelberg, pp 271–287 (data provided on CD-Rom with the book)

Dahlgren RA, Saigusa M, Ugolini FC (2004) The nature, properties and management of volcanic soils. Adv Agron 82:113–182

Einarsson MTh (2013) Effects of soil conservation measures at Rangárvellir on vegetation and carbon fluxes. Unpublished MSc thesis, Agricultural University of Iceland, Hvanneyri

Enkhtuya B, Oskarsson U, Dodd JC, Vosatka M (2003) Inoculation of grass and tree seedlings used for reclaiming eroded areas in Iceland with mycorrhizal fungi. Folia Geobotanica 38:209

Eswaran H, van de Berg E, Reich P (1993) Organic carbon in soils of the world. Soil Sci Soc Am J 57:192–194

Fjellberg A (2007) Icelandic Collembola—revised checklist and general comments. Insect Syst Evol Suppl 64:45–60

Garcia-Rodeja E, Novoa JC, Pontevera X, Martinez-Cortizas A, Buurman P (2007) Aluminium and iron fractionation of European volcanic soils by selective dissolution techniques. In: Arnalds O, Bartoli F, Buurman P, Oskarsson H, Stoops G, Garcia-Rodeja E (eds) Soils of volcanic regions in Europe. Springer, Heidelberg, pp 325–351

Gisladottir G, Erlendsson E, Lal R, Bigham J (2010) Erosional effects on terrestrial resources over the last millennium in Reykjanes, southwest Iceland. Quat Res 73:20–32

Gislason SR (2008) Weathering in Iceland. Jökull—Icel J Earth Sci 58:387–408

Gislason SR (2012) Kolefnishringrásin (the carbon cycle). Hið íslenska bókmenntafélag, Reykjavik (in Icelandic)

Gislason SR, Oekjers EH, Eiriksdottir ES, Kardjilov MI, Gisladottir G, Sigfusson B, Snorrason A, Elefsen SO, Hardardottir J, Torssander P, Oskarsson N (2009) Direct evidence of the feedback between climate and weathering. Earth Planet Sci Lett 277:213–222

Gudleifsson BE (2002) Impact of long term use of fertilizer on surface invertebrates in experimental plots in a permanent hayfield in northern-Iceland. Icel Agric Sci 15:37–49

Gudleifsson BE (2005) Beetle species (Coleoptera) in hayfields and pastures in northern Iceland. Agric Ecosyst Environ 109:181–186

Gudleifsson BE (2007) ICEWOODS: earthworms in Icelandic forest soils. In: Halldorsson G, Oddsdottir E, Eggertsson O (eds) Effects of afforestation on ecosystems, landscape and rural development. TemaNord 2007, pp 127–132

Gudleifsson BE, Olafsson R (1981) Athugun á ánamöðkum í túnum í Eyjafirði (Earthworms in hayfields in Eyjafjordur). Náttúrufræðingurinn 51:105–113 (in Icelandic)

Gudleifsson BE, Bjarnadottir B (2002) List of invertebrates collected in pitfall traps in hayfields and pastures in northern-Iceland 1996–1997. Icel Agric Sci 15:36–37

Gudmundsson B, Thorsteinsson Th (1980) Þungmálmar í íslensku grasi. (Heavy metals in Icelandic grass). Icel Agric Sci 12:3–10 (in Icelandic, English summary)

Gudmundsson Th, Sveinsson Th (2011) Greiningar á jarðvegi tuna og gildi þeirra fyrir ræktendur (Soil analysis and their value for land users). Agric Sci Congr (Fræðaþing landbúnaðarins) 2011:150–159 (in Icelandic)

Gudmundsson J, Oskarsson H (2014) Carbon dioxide emissions from drained organic soils in West-Iceland. In: Halldorsson G, Bampa F, Thorsteinsdottir AB, Sigurdsson BD, Montanarella L, Arnalds A (eds) Soil carbon sequestration: for climate, food security and ecosystem services. JRC scientific and policy records, European Union (in press)

Gudmundsson Th, Björnsson H, Thorvaldsson G (2004) Organic carbon accumulation and pH changes in an Andic Gleysol under a long-term fertilizer experiment in Iceland. Catena 56:213–224

Gudmundsson Th, Björnsson H, Thorvaldsson G (2005) Elemental composition, fractions and balance of nutrients in an Andic Gleysol under a long-term fertilizer experiment in Iceland. Icel Agric Sci 18:21–32

Gudmundsson J, Brink SH, Gisladottir FO (2013) Preparation of LULUCF land-use map for Iceland: development of the Grassland layer and subcategories. Grassl Sci Eur 18:105–107

Guicharnaud R, Arnalds O, Paton GI (2009) The sensitivity of microbial processes in Icelandic soils to increasing temperatures. Biogeosci Discuss 6:6749–6780

Guicharnaud R, Arnalds O, Paton G (2010) The effect of season and management practices on soil microbiological activities undergoing nitrogen treatments—interpretation from microcosm to field scale. Icel Agric Sci 23:123–134

Hallsdottir SH, Wöll C, Gudmundsson H, Snorrason A, Thorsson J (2012) Emissions of greenhouse gases in Iceland from 1990 to 2010. National inventory report 2012. Environmental Agency of Iceland, Reykjavik

Hardarson GH, Thorlacius A, Olafsson BL, Björnsson H, Eiriksson T (2006) Styrkur snefilefna í heyi. (Trace elements in hey). Agricult Congr (Fræðaþing landbúnaðarins) 2006:179–189 (in Icelandic)

Helgason B (2002) Organic phosphorous in Icelandic soils. Icel Agric Sci 15:95–109

Helgadottir A, Eythorsdottir E, Johannesson T (2013) Agriculture in Iceland—a grassland based production. Grassl Sci Eur 18:30–43

Jakobsson SP, Jonasson K, Sigurdsson IA (2008) The three igneous rock series of Iceland. Jökull—Icel J Earth Sci 58:117–138

Johannesson B (1960) The soils of Iceland. University Research Institute, Department of agriculture reports series B, no 13. University of Iceland, Reykjavik Iceland

Johannesson T, Gudmundsdottir KB, Eiriksson T, Jed B, Kristinsson J, Sigurðarson S (2004a) Selenium and GPX activity in blood samples from pregnant and non-pregnant ewes and selenium in hay on scrapie-free, scrapie-prone and scrapie afflicted farms in Iceland. Icel Agric Sci 16–17:3–13

Johannesson T, Gudmundsdottir KB, Eiriksson T, Kristinsson J, Sigurdarson S (2004b) Copper and manganese in hay samples from scrapie-free, scarpie-phrone and scarpie-afflicted farms in Iceland. Icel Agric Sci 16–17:45–52

Johannesson T, Gudmundsdottir KB, Eiriksson T, Kristinsson G, Sigurðarson S (2005) Molybdenum and sulphur in forage samples from scrapie-free, scrapie-prone and scrapie-afflicted farms in Iceland. Icel Agric Sci 18:53–58

Johannesson T, Gugmundsdottir KB, Eiriksson T, Kristinsson J, Sigurdarson S (2007) Overview. Seven trace elements in Icelandic forage. Their value in animal health and with special relation to scrapie. Icel Agric Sci 20:3–24

Lal R (2004) Carbon sequestration in dryland ecosystems. Environ Manage 33:52–544

Madeira M, Gy Füleky, Auxtero E (2007a) Phosphate sorption of European volcanic soils. In: Arnalds O, Bartoli F, Buurman P, Oskarsson H, Stoops G, Garcia-Rodeja E (eds) Soils of volcanic regions in Europe. Springer, Heidelberg, pp 325–351

Madeira M, Auxteros E, Monteiro F, Garcia-Rodeja E, Novoa-Munoz JC (2007b) Exchange complex properties of soils from a range of European volcanic areas. In: Arnalds O, Bartoli F, Buurman P, Oskarsson H, Stoops G, Garcia-Rodeja E (eds) Soils of volcanic regions in Europe. Springer, Heidelberg, pp 369–385

Maljanen M, Sigurdsson BD, Gudmundsson J, Oskarsson H, Huttunen JT, Martikainen PJ (2010) Greenhouse gas balances of managed peatlands in the Nordic countries—present knowledge and gaps. Biogeosciences 7:2711–2738

Nanzyo M, Dahlgren R, Shoji S (1993) Chemical characteristics of volcanic ash soils. In: Shoji S, Nanzy M, Dahlgren RA (eds) Volcanic ash soils. Genesis, properties and utilization. Developments in soil science vol 21. Elsevier, Amsterdam, pp 145–187

Oddsdottir ES, Svavarsdottir K, Halldorsson G (2008) The influence of afforestation and land reclamation on soil arthropods in Iceland. Icel Agric Sci 21:3–13

Oddsdottir ES, Eilenberg J, Sen R, Halldorsson G (2010a) The effects of insect pathogenic soil fungi and ectomycorrhizal inoculation of birch seedlings on the survival of Otiorhynchus larvae. Agric Entomol. doi:10.1111/j.1461-9563.2010.00482.x

Oddsdottir ES, Eilenberg J, Sen R, Harding S, Halldorsson G (2010b) Early reduction of Otiorhynchus spp. larval root herbivory on *Betula pubescens* by beneficial soil fungi. Soil Ecol 45:168–174

Oddsdottir ES, Nielsen C, Sen R, Harding S, Eilenberg J, Halldorsson G (2010c) Distribution patterns of entomophatogenic and brich symbiotic ectomycorrhizal fungi across native woodland and degraded habitats in Iceland. Icel Agric Sci 23:37–49

Oskarsson H (1998) Wetland draining in Western Iceland. In: Olafsson JS (ed) Icelandic wetlands: exploitation and conservation. University of Iceland Press, Reykjavík, pp 121–129 (in Icelandic)

Oskarsson H (2011) Ecological restoration of the wetland committee (Vistheimt á vegum votlendisnefndar). In: Aradottir AL, Halldorsson G (eds) (2011) Ecological restoration in Iceland. Agricultural University of Iceland and Icelandic Soil Conservation Service, Reykjavik, pp 112–127. ISBN 978-9979-881-08-7 (in Icelandic)

Oskarsson H, Brynleifsdottir SJ (2009) The interaction of fertilization in nursery and field on survival, growth and the frost heaving of birch and spruce. Icel Agric Sci 22:59–68

Oskarsson H, Arnalds O, Gudmundsson J, Gudbergsson G (2004) Organic carbon in Icelandic Andosols: geographical variation and impact of erosion. Catena 56:225–238

Oskarsson H, Sigurgeirsson A, Raulund-Rasmussen K (2006) Survival, growth and nutrition of tree seedlings fertilized at panting on Andisol soils of Iceland: six year results. For Ecol Manage 229:88–97

Oskarsson U (2010) Potting substrate and nursery fertilization regime influence mycorrhization and field performance of *Betula pubescens* seedlings. Scand J For Res 25(2):111–117

Oskarsson U (2012) Effectiveness of five fungal isolates as mycorrhizal inoculants of birch seedlings. Icel Agric Sci 25:51–58

Palmason F, Thorgeirsson H, Sigurdardottir H, Björnsson H, Arnalds O (1996) Nitrogen mineralization in soil. Icel Agric Sci 10:185–208 (in Icelandic, English summary)

Panek E, Kepinska B (2002) Trace metal (Cd, Cu, Pb, Zn) and sulphur content in soils and selected plant species of Iceland. A pilot study. Icel Agric Sci 15:3–9

Ping CL, Shoji S, Ito T (1988) Properties and classification of three volcanic ash-derived pedons from Aleutian Islands and Alaska Peninsula, Alaska. Soil Sci Soc Am J 52:455–462

Ping CL, Shoji S, Tto T, Takahasi T, Moore JP (1989) Characteristics and classification of volcanic-ash-derived soils in Alaska. Soil Sci 148:8–28

Ragnarsdottir KV, Hawkins DP (2006) Bioavailable copper and manganese in soils from Iceland and their relationship with scrapie occurrence in sheep. J Geochem Explor 88:228–234

Ritter E (2007) Carbon, nitrogen and phosphorus in volcanic soils following afforestation with native birch (Betula pubescens) and introduced larch (*Larix sibirica*) in Iceland. Plant Soil 295:239–251

Sigurdardottir H (1994) Ánamaðkar (Earthworms). Náttúrufræðingurinn 64:139–148

Sigurdardottir H, Thorvaldsson G (1994) Ánamaðkar (*Lumbricidae*) í sunnlenskum túnum (earthworms in southern hay fields). Icel Agric Sci 8:9–20

Sigurgeirsson MA, Arnalds O, Palsson SE, Howard BH, Gudnason K (2005) Radiocaesium fallout behaviour in volcanic soils in Iceland. J Environ Radioact 79:39–53

Snorrason A, Sigurdsson BD, Gudbergsson G, Svavarsdottir K, Jonsson ThH (2002) Carbon sequestration in forest plantations in Iceland. Icel Agric Sci 15:81–93

Stoops G, Gerrard M (2007) Micromorphology. In: Arnalds O, Bartoli F, Buurman P, Oskarsson H, Stoops G, Garcia-Rodeja E (eds) Soils of volcanic regions in Europe. Springer, Heidelberg, pp 129–140

Stoops G, Gerard M, Arnalds O (2008) A micromorphological study of Andosol genesis in Iceland. In: Kapur S, Mermut A, Stoops G (eds) New trends in micromorphology. Springer, Heidelberg, pp 67–90

Tanneberg H, Jahn R (2007) Heavy metal sorption by andic and non-andic horizons from volcanic parent materials. In: Arnalds O, Bartoli F, Buurman P, Oskarsson H, Stoops G, Garcia-Rodeja E (eds) Soils of volcanic regions in Europe. Springer, Heidelberg, pp 423–435

Thorhallsdottir ThE, Thorsson J, Sigurdardottir S, Svavarsdottir K, Johannsson MH (1998) Disturbance of Wetlands in South Iceland. In: Olafsson JS (ed) Icelandic wetlands: exploitation and conservation. University of Iceland Press, Reykjavík, pp 131–142 (in Icelandic)

Thorvaldsson G, Gudmundsson Th (2006) Snefilefni í jarðrækt (microelements in crop production in Iceland). Agricult Congr (Fræðaþing landbúnaðarins) 2006:170–178 (in Icelandic)

Thorvaldsson G, Gudmundsson Th, Björnsson H (2009) Agricult Congr (Fræðaþing Landbúnaðarins) 2009:224–233 (in Icelandic)

Zehetner F (2010) Does organic carbon sequestration in volcanic soils offset volcanic CO2 emissions. Quat Sci Rev 29:1313–1316

9 Genesis and Mineralogical Characteristics

9.1 Minerals

The understanding of the mineralogy of soils in Iceland remained rather vague until the last part of the twentieth century, as conventional methods for clay research, such as X-ray diffraction, did not show the presence of phyllosilicates. Research by Arnalds (1990, 1994) and Wada et al. (1992) showed that the main mineral constituents consisted of allophane, imogolite, and ferrihydrite. Earlier, Johannesson (1960) makes a brief mention of allophane based on a suggestion of Tu in an appendix of Johannesson's book on the Soils of Iceland. It is also discussed in unpublished reports by Björnsson (1961) and Einarsson (1979). It is interesting to note that F. Weis did try ammonium oxalate extractions of Icelandic soil samples during the 1930s and concluded that weathering rates in Iceland were high, but this comment was disregarded on the grounds that it was too cold in Iceland (cited by Johannesson 1988).

Other minerals than allophane, ferrihydrite, and imogolite occur in Iceland. Gudmundsson (1978, 2009) noted that goethite, siderite, and pyrite occur in some soils, but information on the occurrence of these minerals is otherwise limited. Most attempts to find phyllosilicate minerals (layer silicates) in soils from Iceland in some appreciable amounts have failed. The mineralogical analysis of the clay constituents of the EU COST-622 soil samples did not reveal secondary phyllosilicates: "Practically no reflections of phyllosilicate minerals are visible in XRD diagrams of the <2 μm fraction of the profiles from Iceland" (Monteiro et al. 2007). However, Kleber and Arnalds (unpublished data) found smectites in some samples from the Eastfjords, where the origin may be from sedimentary rocks from the Tertiary basalt stack. Björnsson (1961) also found evidence of smectite based on XRD peaks in samples from alluvial soils in Northwest Iceland and similarly concluded that it originated in Tertiary rocks. He did not find phyllosilicates in adjacent aeolian soils. Smectites and even kaolinites are likely to occur elsewhere in soils under unstable Tertiary basalt stacks with red sedimentary interlayers. However, a suit of secondary clay minerals occur within the numerous thermal areas of Iceland, but they have a limited extent (see, e.g., Gennadiev et al. 2007).

9.1.1 Allophane and Imogolite

The basaltic glass particles have considerable surface area (often >10 $m^2\ g^{-1}$ with broken bonds on the edges). They weather relatively rapidly (see Wolff-Boenisch et al. 2004), releasing basic cations at comparatively fast rate, with oversaturation of Si, Al, and Fe, which form the colloidal constituents allophane, imogolite, and ferrihydrite. Allophane is the chief clay mineral of the soils in Iceland, together with varying amounts of ferrihydrite and less amount of imogolite. Photographs (Transmission Electron Microscopy) of Icelandic allophane were shown in Chap. 5 on Andosols. The idea of allophane was somewhat alien to many naturalists in the beginning, not the least in Iceland, in part because of poor understanding of this mineral early on, with several Icelandic texts rejecting the idea that weathering in Iceland is as rapid as is explained in this chapter, on the basis that it is cold in Iceland and that phyllosilicates are not found in the soils. However, general understanding of the soil genesis is improving with subsequent changes in texts written about Icelandic nature.

How much allophane is there in Icelandic soils? The amount varies of course, but generally ranges between 2 and 15 %, but higher contents do occur. The total content is dependent on the weathering environment. Allophane content is low in the desert Vitrisols (see Sect. 6.5), often 2–5 %, which are youthful or continuously disturbed soils, not allowing for rapid allophane accumulation, although weathering rates can be considered rather high. Allophane content is also low in organic soils with low pH, far from aeolian sources. This is because allophane formation is inhibited at pH below about 4. 9, with Al–humus complex formation becoming dominant as the pH becomes lower. Icelandic Andosol surface samples analyzed by Sigurgeirsson et al. (2005) provide excellent evidence of this, as is shown in Fig. 9.1.

O. Arnalds, *The Soils of Iceland*,
World Soils Book Series, DOI 10.1007/978-94-017-9621-7_9
© Springer Science+Business Media Dordrecht 2015

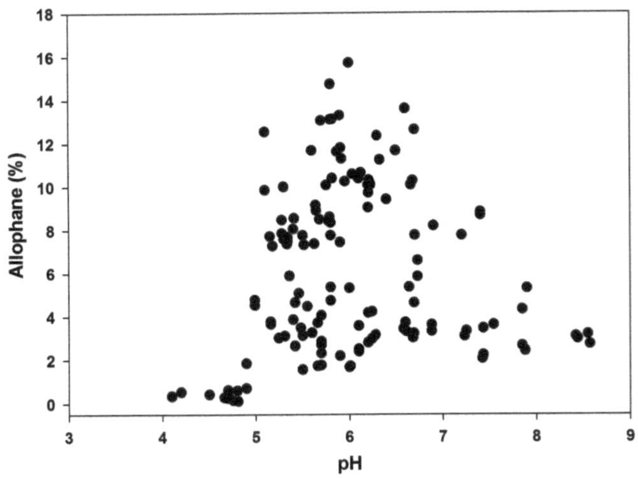

Fig. 9.1 Allophane and pH (measured in H_2O) in surface samples. Samples obtained from 38 sites at 31 locations are widely distributed over the country. Top 0–15 cm were sampled, obtained from 2 or 3 depth intervals. Allophane content is abruptly cut near pH 4.9. Soils with the highest pH are vitric in nature with low allophane content. The samples were obtained for ^{137}Cs research, see Sigurgeirsson et al. (2005)

One of the characteristics of allophane is that it does not have quite fixed chemical composition, with variable Al/Si ratio. This molar ratio is obtained by extracting the soils with acid oxalate and pyrophosphate, the molar ratio calculated as $(Al_{ox} - Al_{pyr})/Si_{ox}$. The pyrophosphate extractable Al is associated with organic complexes (see Garcia-Rodeja et al. 2007). The AUI soil database shows clear relationship between Al_{pyr} and total organic content, with more aluminum–humus complexes occurring in the organic soils with lower pH. The Al/Si ratios of the Icelandic allophanes (and imogolite, combined in the bulk soil) range generally between 0.8 and 2 (AUI soil database), with vitric soils having this ratio near 1, and allophanes in andic soil horizons having a cluster around 1.2, however, with a large variability. Meijer et al. (2007) reported Al/Si ratios of 0.9–1.1 (one with 1.5) for peons from three EU COST profiles in Iceland. These values represent a lower Al/Si range than commonly found for the world's Andosols, which is most likely attributed to the basaltic nature and the rapid weathering of the parent materials. All Icelandic Andosols have vitric character and are less weathered than the average Andosols of Europe (Taboada et al. 2007) and the molar ratio of allophane in these soils are similar to the more vitric horizons in the EU COST-622 database (Garcia-Rodeja et al. 2007). The formation and structure of allophanes with these low Al/Si ratios are less understood than the more common allophanes with Al/Si near 2 (Dahlgren et al. 1993). Some of the oxalate extractable Si (Si_{ox}) can in fact be chemisorbed to the surfaces of the ferrihydrite, amplifying the low (Si rich) Al/Si ratios (see Arnalds et al. 1995). The fate of the Si depends on the Al availability and Opfergelt et al. (2011) suggested that opaline silica was likely to form in the Si rich Icelandic weathering environment, but opaline silica is common in many other Andosols, such as in Japan (Dahlgren et al. 1993). However, the author is not aware of available information about opal in soils in Iceland.

Imogolite was found in some but not all horizons in study by Wada et al. (1992). It can be concluded that allophane is the dominant mineral together with ferrihydrite, and thus imogolite is not considered in general discussions of the mineralogy below.

9.1.2 Ferrihydrite

Ferrihydrite is common in young iron oxide accumulations and as bog iron consisting of small (3–7 nm) poorly ordered Fe^{3+} (Schwertmann and Taylor 1989; Bigham et al. 2002). Ferrihydrite is most commonly calculated as oxalate extractable iron (Fe_{ox}) times the factor 1.7 (Parfitt and Childs 1988). Ferrihydrite is generally 2–10 % in soils in Iceland, although horizons with up to 20 % ferrihydrite exist. The AUI soil database (using >400 horizons) showed no clear relationship between ferrihydrite and pH (in H_2O) or organic carbon. The highest ferrihydrite contents (>10 %) are generally associated with subsurface horizons, but not exclusively. These higher contents were not associated with particular types of soils or regions, but are likely to reflect reduction–oxidation environment of each profile, which is influenced by freeze/thaw process, water retention, coarse grained tephra layers and other factors. Garcia-Rodeja et al. (2007) noted that high Fe values in some horizons were likely to occur "due to gleying upon water saturation caused by frost blockage and/or freezing thawing processes."

Some of the ferrihydrite can be in early stages of goethite formation, which was observed under the microscope by Stoops et al. (2008).

9.1.3 Organo-mineral Complexes

Using many different dissolution technics can yield important information about the mineral constituents of soils and these methods have widely been employed on Andosols, especially since conventional methods such as XRD are often of limited use (e.g., Garcia-Rodeja et al. 2007). Pyrophosphate extractions (Al_{pyr} and Fe_{pyr}) reveal the amount of Al and Fe in the form of organo-mineral complexes. The AUI soil database shows that the amount of Al and Fe bound by organo-mineral complexes (allophane, ferrihydrite, and Al/Fe humus complexes) is closely correlated with the total carbon content (Fig. 9.2), with an increasing proportion of the $(Al + Fe)_{ox}$ as pyrophosphate extractable Al + Fe ($R^2 = 0.77$; $n = 77$).

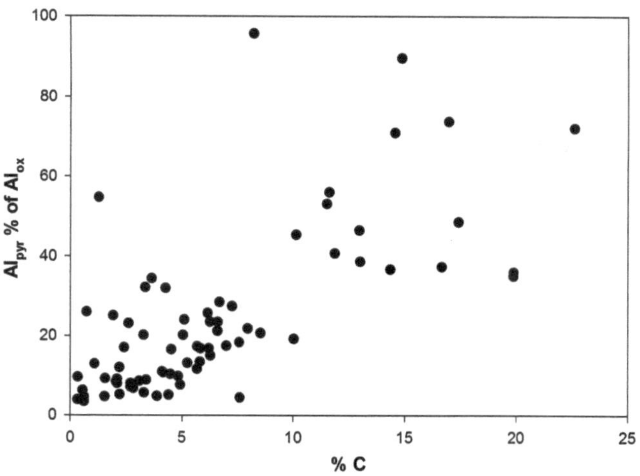

Fig. 9.2 Pyrophosphate extractable Al (Al_{pyr}) as a proportion (%) of total extractable Al (Al_{ox}), indicating how much of the extractable Al is associated with aluminum–organo complexes. Higher proportion of the Al is bound to organo-mineral complexes as the total organic content of the horizons increases. Based on AUI soil data

Common ratios of (Al + Fe)/C in organo-mineral complexes in Andosols range between 0.1 and 0.25 (Dahlgren et al. 1993; Nanzyo et al. 1993; Rodriguez et al. 2006). Using this ratio reveals that a significant proportions (often 20 to >80 %) of the carbon in the soil is associated with organo-mineral complexes, but this awaits further studies (see Arnalds et al. 2013). It is noteworthy that the AUI soil database and other studies (Arnalds et al. 1995, 2013) show considerable amounts of organo-mineral complexes at relatively high pH. Garcia-Rodeja et al. (2007) had indicated that metal–humus complexion would be limited at pH (H_2O) > 6, but the complexion at the higher pH may be primarily associated with allophane and ferrihydrite surfaces.

9.2 Total Chemical Composition

The mineralogy of soils of Iceland is dominated by the volcanic origin of the parent materials, with varying amounts of secondary minerals characteristic of Andosols (see Chap. 5). As most of the parent materials originate in aeolian dust sources, the total chemical composition is influenced by the geochemistry of these sources. However, the composition may be compounded by redistribution of soils caused by wind erosion of already weathered soil materials. Oskarsson et al. (2012) published a comprehensive table showing the composition for the major elements for 62 horizons from 16 Andosol pedons in Iceland (soils under vegetation). These horizons were at various stages of weathering. The SiO_2 content of these horizons ranged from about 29 to 61 %, with about half of the horizons with <40 % SiO_2. Al_2O_3 generally ranges between 13 and 17 %. Iron content shows more variability, with Fe_2O_3 (total Fe) ranging from 3.5 to >17 %, with lower values caused by both differences in parent materials (rhyolite inputs) and reduction and leaching of iron. A graph showing major element composition in selected profiles is presented in Fig. 9.3 (from Oskarsson et al. 2012).

Figure 9.3 clearly demonstrates the erratic distribution of the major elements with depth, which is influenced by the various inputs of aeolian and tephra materials. An example of this is the AR pedon (lower left), a pedon from South Iceland relatively close to Mt. Hekla (note the many tephra layers), but this soil has also received major additions of redistributed soils from the highland areas close by, in part rich in silicious tephra. The thick tephra layer at the MV site is silicious Hekla tephra, note the increase in SiO_2. Chemical reactivity and subsequent mobility is most active near the surface, but slows down as the materials become buried under more recent materials.

Oskarsson et al. (2012) found that Ti, Al, Fe, and Mn were the least mobile elements and thus often found enriched within mature horizons. Mg, Ca, and Na are depleted as a result of pedogenesis.

Volcanic glass is the main constituent of the sand fraction of most of the Vitrisols (desert soils, classify as Vitricryands according to ST) and Andosols. The character of the sand fraction varies considerably in the Andosols, representing the aeolian and tephra deposition environment, with strong stratification (Fig. 9.4). Tephra is often a dominating component, together with other pyroclasts, but rock fragments of augite, feldspar, and olivine also occur as expressed by 11 horizons from three pedons of the EU-COST-622 project (see Stoops and Van Driessche 2007). Rock fragments are often more abundant in the lower parts of the profile, representing lithological discontinuities, i.e., change in the aeolian environment. Glacial till C horizons often underlie the Andosols with contrasting mineralogy, chiefly made of basaltic rock fragments (see also Arnalds et al. 1995).

The mineralogy of the Vitrisols is generally dominated by the geochemistry of the volcanic system influencing the desert surfaces, with the Katla volcano dominating the Mýrdalssandur, Grímsvötn volcanic system controlling the Skeiðarársandur, and Bárðarbunga and Kverkfjöll volcanic systems dominating the Northeast desert sands (see, e.g., Baratoux et al. 2011). Some desert sands consist primarily of rock fragments, created by glaciers advancing over and reworking lava surfaces, such at the southern margin of Langjökull glacier (Mangold et al. 2011). The Cambic Vitrisols, characteristic of the lag gravel surfaces, often have a mixture of andic materials, vitric and rock fragments, expressing a very weakly developed Bw horizon.

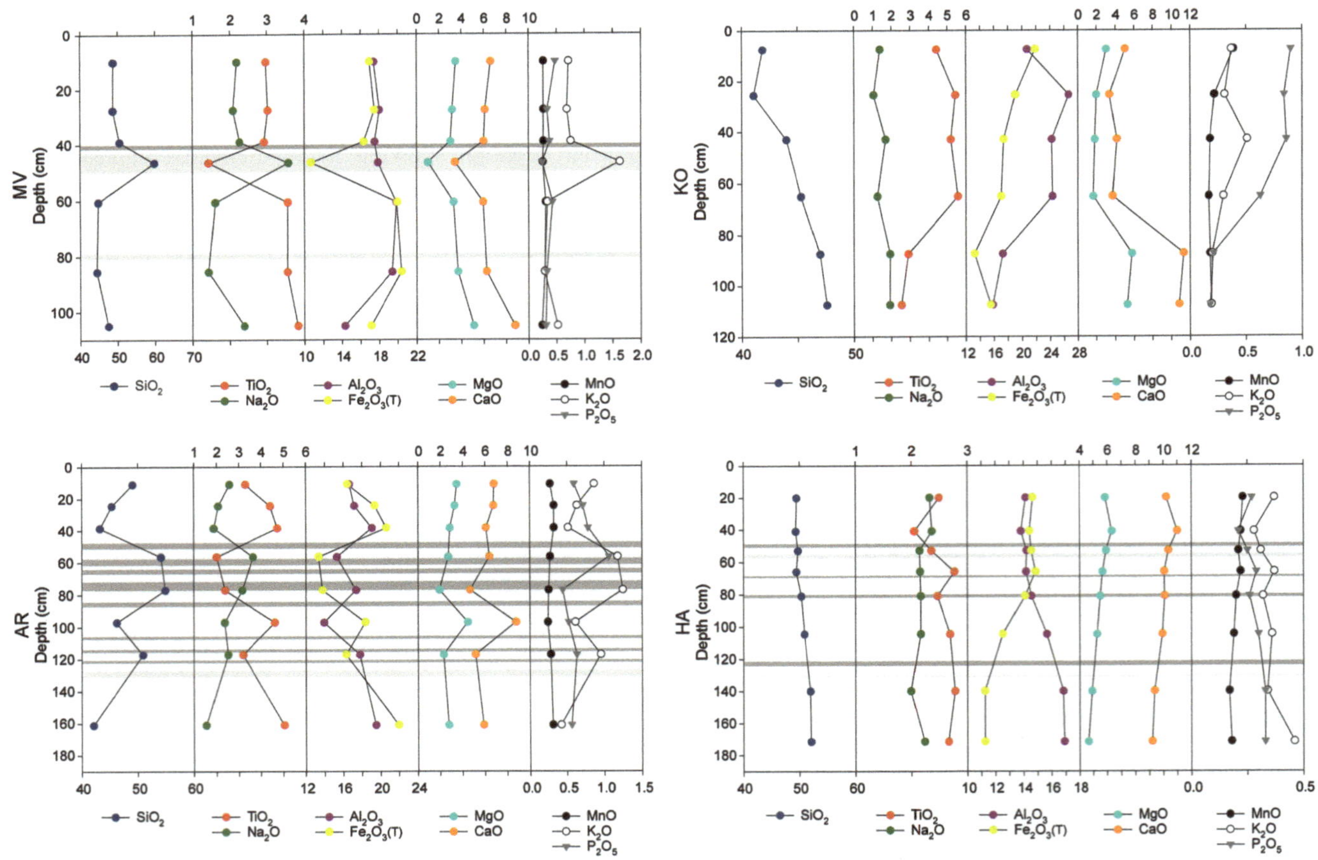

Fig. 9.3 Major element composition in four selected profiles. From Oskarsson et al. (2012). *Shaded horizontal lines* represent major tephra layers. Locations are: *MV* Möðruvellir, North Iceland; *KO* Korpa, Southwest Iceland; *AR* Árnes, South Iceland; *HA* Hálslón, Eastern Highlands

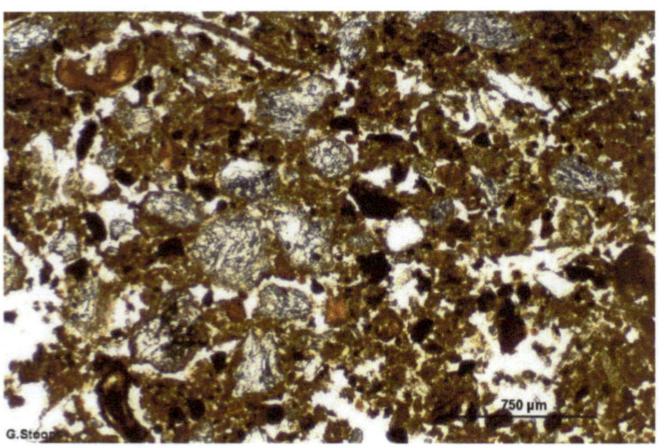

Fig. 9.4 Thin section micrograph from the surface horizon of the COST-622 09 (Auðkúluheiði) pedon. Thin diffuse coatings of fine material around pumice fragments, and partially penetrating in the open vacuoles. Frequent fragments of brown partially decomposed organic material and frequent coarse sand and gravel size grains of pumice with fine material coatings. Occasional basalt fragments e.g., olivine. From Stoops et al. (2007); see also Stoops et al. (2008). *Photo* Georges Stoops

9.3 Micromorphology

Micromorphological studies of Icelandic soils are relatively few. Early attempts include those Gudmundsson (1978) who studied peat soils in Northwest Iceland. He found that the structure of the roots and rhizomes were preserved in his wetland pedons, with relatively little disturbance by animals. He identified diatoms near mineral layers, siderite in lenses and also pyrite in the lower parts, and concluded that chemical weathering and secondary mineral formation were active in the peats he studied. Other studies include those of Romans et al. (1980), who studied recently exposed glacial tills in front of the retreating Breiðamerkurjökull glacier (South Iceland) and Simpson et al. (1999), who used micromorphology as a tool in archeological research near Lake Mývatn, in Northeast Iceland. A considerable effort was made to characterize mineralogical and micromorphological characteristics of volcanic soils in Europe during the COST-622 project, which resulted in numerous publications involving soils from Iceland (three pedons), such as Stoops and Gerard (2007), Stoops

Fig. 9.5 Colors in part dominated by secondary minerals in hydrothermally altered interior of the Torfajökull caldera, South Central Iceland

and Van Driessche (2007), Meijer et al. (2007), Monteiro et al. (2007), Bartoli and Burtin (2007), and Basile et al. (2007), which were in part summarized in the book chapter "A micromorphological study of Andosols Genesis in Iceland," by Stoops et al. (2008). Van Vliet-Lanoe and coworkers have studied micromorphology in Iceland in relation to frost action (e.g., Van Vliet-Lanoe et al. 1998). Hydrothermally affected areas have totally different mineralogy and morphology from other areas (Fig. 9.5).

Judging from the literature, the mircromorphology of soils in Iceland (Figs. 9.6, 9.7, 9.8, and 9.9) is in general characterized of various combinations of (i) poorly weathered vitric materials, which both include basaltic (sideromelane and tachylite) and rhyolitic materials together with palagonite fragments; (ii) rock fragments, with olivine, pyroxene, and plagioclase as the most common minerals; (iii) various organic matter fragments of different degree of decomposition; and (iv) various secondary amorphous mineral products, which are most commonly allophane and ferrihydrite. The common aeolian and tephra deposition events result in

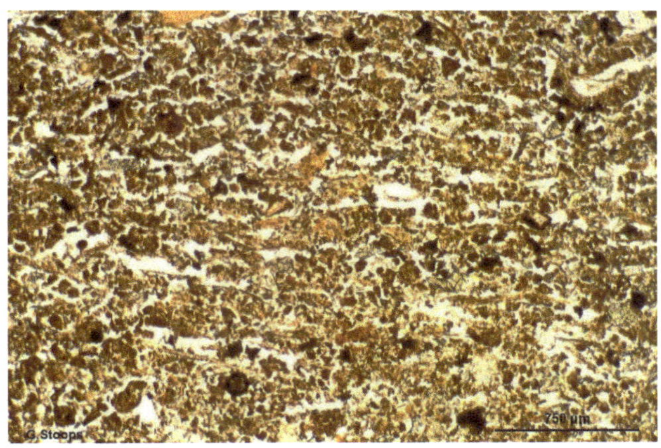

Fig. 9.6 Micrograph of the 4Bw horizon of the Ós pedon (COST 622 07). Angual grains of augite, feldspar and rounded grains of green glass. Black vesicular pyroclasts and aggregates of chlorite. Few rounded feldspathic pyroclasts with chlorite inclusions. The most striking feature of this picture is the lamellar, partly lenticular microstructure, pointing to an alternation of freezing and thawing. From Stoops et al. (2007); see also Stoops et al. (2008). *Photo* Georges Stoops

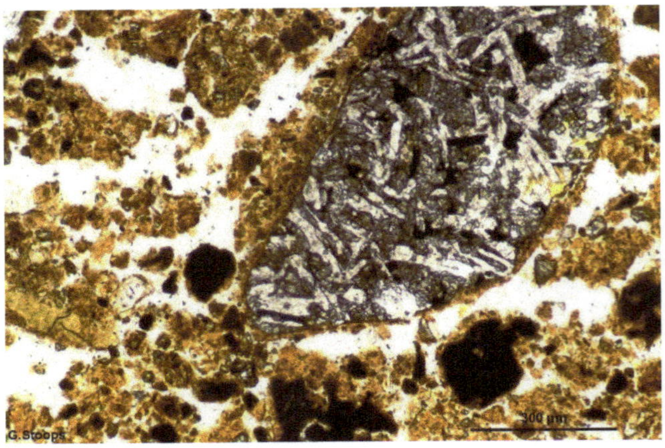

Fig. 9.7 Micrograph of the 2CB horizon at Ós (COST 622 07). Mineral fragments of brown greenish glass with few angular grains of feldspar and augite. Diatoms are abundant. Reddish brown organic tissue residues. Subrounded holocrystalline volcanic rock fragment and smaller angular glass fragments; partly granular microstructure, characteristic for andic materials. From Stoops et al. (2007); see also Stoops et al. (2008). *Photo* Georges Stoops

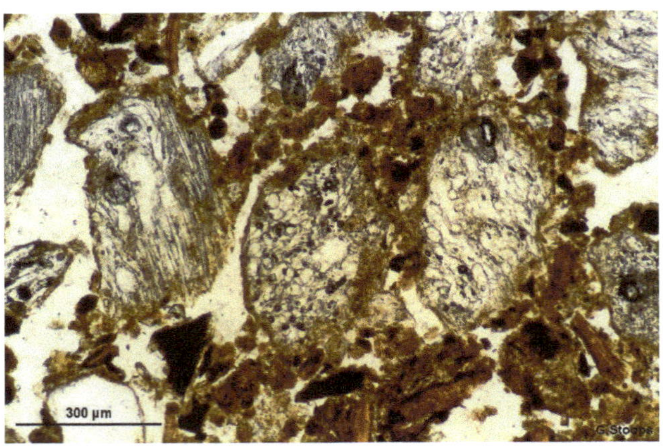

Fig. 9.8 Micrograph of Ah2 horizon from Auðkúluheidi (COST 622 08). Subrounded pumice fragments (about 500 μm) covered by coatings of fine material penetrating in the elongated vacuoles at the outer rim (internal clay hypocoatings); subangular grains of green volcanic glass. From Stoops et al. (2007); see also Stoops et al. (2008). *Photo* Georges Stoops

multiple lithological discontinuities, showing microstratification that is evident in thin sections, both of mineral and organic horizons. The rhyolite tephra is especially resistant to weathering and often has limited weathering fringes. However, Stoops et al. (2008; Stoops personal communication) points out that in coarse soils, the >2 mm tephra grains are sieved out before analysis, but these grains commonly have a large quantity of clay and amorphous material present in the open vacuoles (hence the high surface area of the vitric materials mentioned above). This means that the physical and chemical analysis (e.g., CEC, clay content, base saturation, etc.) are not correct, but the extent of this influence of coarse materials has not been investigated.

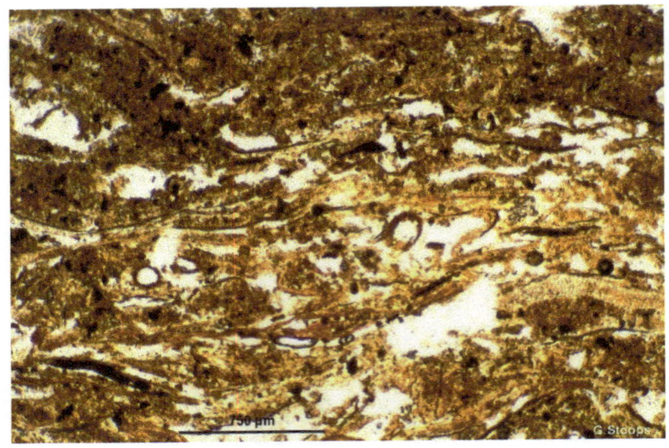

Fig. 9.9 A micrograph from the Ah horizon of the Hella pedon (COST 622 09). Some grayish pumice fragments are present, some angular augite grains and greenish glass. Elongated phytoliths and diatoms are common. The parallel orientation of the organic matter is not horizontal. From Stoops et al. (2007); see also Stoops et al. (2008). *Photo* Georges Stoops

Some horizons show signs of iron cementation (Arnalds et al. 1995) and even early stages of goethite transformation of ferrihydrite in iron-rich horizons (Stoops et al. 2008). Some early attempts on identifying the weathering products in the soils may have used the term móberg (palagonite) for poorly ordered secondary minerals amorphous under the microscope. Freeze–thaw phenomena shows up in the microfabric (Stoops et al. 2008; Van Vliet-Lanoe et al. 1998). The limited weathering of some of the mineral matter may suggest that most of these were deposited directly as ash, but one has also to consider that much of the aeolian sources consist of sand flats with relatively unweathered vitric materials (Baratoux et al. 2011).

Douglas (1987) noted manganese-rich coatings on Icelandic rock surfaces, especially in fractures of basalt lavas. These are likely to affect the color of the dark Icelandic deserts.

9.4 Genesis

Björn Johannesson's book "The Soils of Iceland" from 1960 contains discussions on soil genesis, noting the importance of the basaltic nature of the rocks and the aeolian influence ("loessial soil"). He noted the difference in deposition rates (soil thickening rates) before and after the Settlement, with about tenfold increase in East Iceland, citing the work of the geologist Sigurdur Thorarinsson. He also noted the importance of differences in the drainage of the bedrock, resulting in the division of wetlands and dryland soils. Other early publications on the genesis of soils in Iceland include those of Helgason (1968) who studied soils of Southwest Iceland. Many early genesis studies were devoted to soil thickening rates, which in Icelandic context was sometimes considered "soil formation" (see Thorarinsson 1961; Gudbergsson 1975). Sigbjarnarson (1969) concluded after studying thickening rates near one of the most active dust sources in Iceland that all majority of the mineral parent materials were basaltic volcanic glass originating from the distant desert areas (volcanic loess), but his conclusions met with skepticism initially, with many considering that more of the materials were locally derived and consisting of palagonite. However, the sources of these volcanic loess materials remained rather unclear, but have now been placed at the very active dust plume areas on one hand and the sandy desert areas in general on the other (Arnalds 2010), with contributions from erosion and redistribution of existing Andosols in variable amounts, as is discussed in the chapter on the aeolian environment (Chaps. 11 and 12).

9.4.1 Andosols

The genetic pathways of the Icelandic soil formation vary with soil type, ranging from organic Histosols to the unstable barren desert surfaces. All these soils are subjected to chemical weathering, with the formation of colloidal constituents (clays and organo-mineral complexes). The aeolian additions (and tephra deposition), with the rising soil surface, is a major characteristics of all soils with stable surfaces—but to a varying degree, depending on the rate of sedimentation. Carbon accumulation is also typical of stable surfaces, affected by aeolian characteristics and drainage. Sigfusson et al. (2008) and Oskarsson et al. (2012) showed that chemical weathering is greatest, while the parent materials are close to the surface, but slows down as they become buried. Chemical weathering is most effective in the biologically active surface layer, and the degree of chemical weathering for each horizon is therefore in part affected by how long each horizon remains near the surface before it becomes buried under new aeolian accumulations. This generates an interesting interaction between chemical weathering and aeolian burial, where one has to consider both the characteristics of each horizon and the total thickness of the profile. The fast weathering of basalt controls the chemistry of the soil water on one hand, and the precipitation of allophane and ferrihydrite on the other (see Sigfusson 2004). These fast rates are well expressed both by rapid geochemical denudation (Gislason 2008; Gislason et al. 2009) and the fast rate of allophane formation discussed below. Yet, each horizon consists of relatively little weathered materials as the soil is subjected to these rapid rates of weathering over a short time (decades, hundreds of years).

How much aeolian materials are deposited? How fast are the surfaces rising? Arnalds (2010) reviewed literature on thickening rates and found that the surfaces are rising on average from <0.01 mm far from aeolian sources to as fast as >1 mm each year within the most active aeolian areas. Research that utilizes thickening rates of aeolian deposits includes the classical paper by Thorarinsson (1961) on wind erosion in Iceland, which covered much of Iceland, Gudbergsson (1975, 1996) in the Skagafjörður area in the central North, Sigbjarnarson (1969) in the South, and Gisladottir et al. (2010) in the Southwest. More papers of this nature will be discussed in the Chap. 12. The burial/additions characteristics of the soils are illustrated in Fig. 9.10.

Figure 9.10 is a drawing of the development of a Brown Andosol located near Goðafoss in Northeast Iceland (descriptions and data for this pedon are presented in Chaps. 7 and 8). This pedon is selected for illustration because it is located within area that is well vegetated (the pedon representing regional rather than local changes in aeolian activity), large amount of analytical data that is available, and it has easily identifiable and dated tephra layers. Soil development begins after the glacier retreats, and vegetation gets established on the surface. Aeolian materials accumulate gradually on the surface, building an A horizon, which again becomes slowly buried, however, with continued soil development, to become a Bw (or 2Bwb) horizon. This steady burial continues, but is disrupted by the deposition of thick rhyolitic tephra deposits from Mt. Hekla about 4,250 and 3,150 years ago (from 220 km distance). There is a marked change in the soil development about 1,100 years ago, when there is a large increase in aeolian deposition rates, and the soils are noticeably lighter above this color change in the profile. This change is brought about by the Settlement. The reason for the lighter color is in part change in organic content and a larger proportion of yellowish colored tephra grains which are being redistributed by aeolian processes due to massive soil erosion upwind. These rhyolitic grains originate from the H3 and H4 tephra layers in soils further south and are being redistributed toward north during dry southerly high intensity wind events.

The change in aeolian deposition rates has an influence on carbon accumulation and clay formation. The Goðafoss pedon presented here in Fig. 9.10 was divided into five periods based on the age of the major tephra layers present in the profile. The average thickening rate before the settlement ranged between 0.048 and 0.11 mm $year^{-1}$ but has been 0.51 mm $year^{-1}$ over the past 500 years (Table 9.1). This is based on present bulk thicknesses, not considering losses by chemical weathering and gains of organic carbon (see note on aeolian thickening rates in Chap. 12). Carbon has been accumulating much faster during the past 500 years than before the Settlement when aeolian deposition rates were considerably slower. The same applies for the clay (allophane + ferrihydrite), with about 60 g of clays forming each year under a square meter at present time. The average Al_2O_3 content of the Vatnajökull volcanic sources is about

Fig. 9.10 Genesis of Icelandic soil under a steady accumulation of aeolian materials and periodic ash-fall events. The graph shows 9–10,000 years evolution, starting with a glacial till surface (*left*), which gradually is buried under more and more of aeolian materials, however with continuous soil development, with losses of cations but gains of organic materials and clay constituents. Graph based on Arnalds (1990) and Arnalds et al. (1995)

14 % (Oladottir et al. 2011), and the Al is relatively immobile in the soils (Oskarsson et al. 2012). Dissolution of 9–140 tons km^{-2} year^{-1} of rocks (range in clay formation in Table 9.1) is needed for the formation of the allophane clays alone based on the Al content. This range is similar as is reported by Gislason (2008) for chemical denudation rates for Iceland (see below).

The main pattern of the chemical weathering of the Icelandic Andosols is characterized by the accumulation of Al and Fe with minimal translocation (Oskarsson et al. 2012) as is characteristic of Andosols, together with accumulation of organic materials (Dahlgren et al. 1993). However, Oskarsson et al. (2012) also found evidence of mobilization of Al/Fe complexes resembling podsolization, which also occurs in other volcanic areas (Ugolini and Dahlgren 2002). Guicharnaud and Paton (2006) showed that eathering rates in the Icelandic Andosols were considerable faster than in Cambisols from Scotland from similar climatic region, and had a much greater buffering capacity with stable pH under experimental acid rain deposition.

9.4.2 Vitrisols—The Vitric Soils of the Deserts

The most comprehensive overview of the Icelandic Vitrisols was published by Arnalds and Kimble (2001). The Vitrisols meet the criteria for Andosols (WRB) and Andisols (Soil Taxonomy) by consisting of volcanic glass and having $(Al + ½Fe)_{ox} > 0.4$ %, except where WRB depth requirements are not met (no such requirements in Soil Taxonomy). The oxalate extractable Al and Fe are thought of as being the result of pedogenesis according to those classification systems. It is therefore interesting to note that many of the sand flats classifying as Andosols according to both WRB and Soil Taxonomy have surface materials deposited in recent floods associated with volcanic eruptions, such as the 1918 flood at Mýrdalssandur and the 1996 at Skeiðarársandur. Some of these basaltic glass materials have developed enough "allophanic like" materials (oxalate extractable) upon deposition of the tephra materials. However, limited extent of weathering is still the very character of the Vitrisols. They are characterized by low organic matter and clay contents compared to the Brown Andosols. The allophanes, often 1–5 % have very low Al/Si ratio (often <1) characteristic of vitric materials elsewhere (see section on allophane above). The soils are young, but the weathering rates of these materials are relatively rapid, even though they are lacking mostly the biological activity characteristic of the soils under vegetation in Iceland.

It should be noted that the desert Vitrisols of Iceland are quite variable, mostly depending on the geomorphic surfaces, ranging from sandy materials, gravelly glacial till (lag gravel) to scree and lava surfaces (see Chap. 6 on classification).

Table 9.1 Soil thickening rates, carbon and clay accumulation in the Goðafoss pedon between tephra layers of known age, based on Arnalds (1990)

Period	Time interval	Thickening rate[a] (mm $year^{-1}$)	Carbon (g m^{-2} $year^{-1}$)	Clay (g m^{-2} $year^{-1}$)
"a"–present	1480–1987 AD	0.51	17.9	62.3
Settlement–"a"	874–1480 AD	0.19	8.4	9.4
H3–Settlement	3150 PB–874 AD	0.039	1.2	6.5
H4–H3	4250–3150 BP	0.12	2.3	16.0
Till–H4	9000–4250 BP	0.059	1.2	10.2

Dates for Hekla (H) tephra layers from Larsen and Eiríksson (2008)

[a] Based on present bulk thickness, not considering losses by chemical weathering, organic additions and changes in particle arrangements

9.4.3 Histosols

There are two main studies of the genesis of Histosols in Iceland, firstly by Gudmundsson (1978) and secondly research of two wetland pedons (not Histosols per se) by the European COST-622 group (e.g., book edited by Arnalds et al. 2007). The main characteristic of the genesis of the Histosols is, of course, the accumulation of organic matter. Gudmundsson (1978) noted similar rates of organic accumulation (0.1–0.45 mm $year^{-1}$) in Iceland as reported for peat soils in neighboring countries. However, the accumulation occurs under various rates of aeolian accumulation and periodic tephra deposition events. The cold climate results in reduced decomposition rates, leading to the accumulation of poorly decomposed organic matter. Unique feature of the Icelandic Histosols is the sporadic occurrence of tephra layers, which allows for calculations of organic accumulation rates. The chief difference of the wetland soils in general are the variable organic contents and the degree of humification.

9.5 Chemical Weathering and Denudation

Rapid weathering of volcanic tephra has been emphasized as an important feature of the formation of Andosols and the colloids that make up the mineral constituent of these soils (see Chap. 5). As Iceland is located in a cold climate, one may expect that the weathering rates are not especially rapid and many Icelandic naturalists believed that chemical weathering in Iceland was slow or nonexisting until recently. Other factors than the climate are also important, of course, such as the crystallinity and the composition of the volcanic materials. But how rapid is chemical weathering in Iceland? The short answer: it is very rapid on a global scale. The numbers above indicating the rate of clay formation and how much rock dissolution was needed (9–140 tons km^2 $year^{-1}$) clearly indicate that the rate is high. Some of the other elements (other than Al) dissolved during the soil genesis are in part incorporated into clay minerals such as iron in ferrihydrite and silica in allophane, but some are relatively immobile such as titanium.

Sigurdur R. Gislason and coworkers have studied the chemical weathering rates and chemical denudation in various parts of Iceland (Eiriksdottir et al. 2008; Gislason et al. 2006, 2009; Kardjilov et al. 2006), which was reviewed by Gislason in the journal Jökull special issue on Icelandic geology in 2008. He concluded that chemical denudation rates in Iceland, of the order of 20–150 tons km^{-2} $year^{-1}$ were 1.3 times the world average, which occurs in spite of the cold climate. These rates are very comparable to the rock dissolution rates calculated for allophane clay formation above. The high weathering rates are especially interesting as chemical denudation rates are generally highest for carbonate dominated areas ($CaCO_3$). Poorly crystalline olivine, with many broken edges and bonds, is a major constituent of the glass that dominates the parent materials of the soils in Iceland and it weathers rapidly. Poorly crystalline plagioclase and pyroxenes are also major components of the glass and they also weather relatively rapidly. The olivine and plagioclase minerals are calcium bearing, and the Ca^{++} is lost from the soil profile during weathering and it eventually reacts with CO_2 to form carbonate ($CaCO_3$) in the oceans or the bedrock. Thus, large quantities of CO_2 is drawn from the atmosphere with chemical weathering in Iceland, at a rate that will increase with global warming, thus having a counterbalancing effect on climate change (Gislason 2008; Gislason et al. 2009). The weathering rates vary between the geological formations. Holocene lavas have high permeability which allows for ready infiltration and limited surface runoff, at least during summer (see Chap. 10 on Cryoturbation). The Quaternary and Tertiary formations have slower permeability, resulting in more surface runoff. The average water runoff rates are 1,460 mm $year^{-1}$, an order of magnitude higher than the average world runoff rates of 372 mm $year^{-1}$ (Jonsdottir 2008).

The chemical denudation rates as summarized by Gislason (2008) were highest for the younger geologic formations, which incidentally, are also areas of highest aeolian sedimentation rates. Climatic factors, which are affected by elevation, are important. Eiriksdottir et al. (2013) concluded that chemical denudation increased by 13 % with each 1 °C increase in temperature, but that it was also influenced by runoff rates. The most mobile elements are F, S, Na, K, Ca, and Mg (see also Oskarsson et al. 2012). Aluminum is not mobile and precipitates mostly as allophane (and some imogolite) together with part of the silica.

References

Regarding punctuation and Icelandic characters in citations: See section on punctuation in the Preface

Arnalds O (1990) Characterization and erosion of Andisols in Iceland. Unpublished PhD dissertation, Texas A&M University, College Station, Texas

Arnalds O (1994) Leir í íslenskum jarðvegi (clays of Icelandic soils). Náttúrufræðingurinn 63:73–85 (in Icelandic, extended English summary)

Arnalds O (2010) Dust sources and deposition of aeolian materials in Iceland. Icel Agric Sci 23:3–21

Arnalds O, Kimble J (2001) Andisols of deserts in Iceland. Soil Sci Soc Am J 65:1778–1786

Arnalds O, Hallmark CT, Wilding LP (1995) Andisols from four different regions of Iceland. Soil Sci Soc Am J 59:161–169

Arnalds O, Bartoli F, Buurman P, Oskarsson H, Stoops G, Garcia-Rodeja E (eds) (2007) Soils of volcanic regions in Europe. Springer, Heidelberg

Arnalds O, Orradottir B, Aradottir A (2013) Carbon accumulation in Icelandic desert Andosols during early stages of restoration. Geoderma 193–194:172–179

Baratoux D, Mangold N, Arnalds O, Bardintzeff J-M, Platevoet B, Grégorie M, Pinet P (2011) Volcanic sands of Iceland—diverse origins of aeolian sand deposits revealed at Dyngjusandur and Lambahraun. Earth Surf Process Landf 36:1789–1808

Bartoli F, Burtin G (2007) Organo-mineral clay and physical properties in COST-622 European volcanic soils. In: Arnalds O, Bartoli F, Buurman P, Oskarsson H, Stoops G, Garcíi-Rodeja E (eds) Soils of volcanic regions in Europe. Springer, Heidelberg, pp 469–491

Basile A, Coppola A, De Mascellis R, Mele G, Terribile F (2007) A comparative analysis of the pore system in volcanic soils by means of water-retention measurements and image analysis. In: Arnalds O, Bartoli F, Buurman P, Oskarsson H, Stoops G, Garcia-Rodeja E (eds) Soils of volcanic regions in Europe. Springer, Heidelberg, pp 515–537

Bigham JM, Fitzpatrick RW, Schulze DG (2002). Iron oxides. In: Dixon JBG, Schulze DG (eds) Soil mineralogy with environmental applications. Soil Science Society of America book series, vol 7. Soil Science Society of America, Madison, pp 323–366

Björnsson H (1961) Analyser på isländska jordar. Unpublished report, Agricultural University of Iceland

Dahlgren R, Shoji S, Nanzyo M (1993) Mineralogical characteristics of volcanic ash soils. In: Shoji S, Nanzyo M, Dahlgren R (eds) Volcanic ash soils—genesis, properties and utilization. Elsevier, Amsterdam, pp 101–143

Douglas GR (1987) Manganese-rich rock coatings from Iceland. Earth Surf Proc Land 12:301–310

Einarsson S (1979) Allofaner i eoliske sedimenter og tefraavsetninger på Island. Hovedopgave (thesis), Norges landbrukshögskole, Ås, Norway, unpublished (in Norwegian)

Eiriksdottir ES, Louvat P, Gislason SR, Oskarsson NÖ, Hardardottir J (2008) Temporal variation of the chemical and mechanical weathering in NE-Iceland, evaluation of a steady state model of erosion. Earth Planet Sci Lett 272:78–88

Eiriksdottir ES, Gislason SR, Oelkers EH (2013) Does temperature or runoff control the feedback between denudation and climate? Insights from NE Iceland. Geochimica et Cosmoshimica Acta 107:65–81

Garcia-Rodeja E, Novoa JC, Pontevera X, Martinez-Cortizas A, Buurman P (2007) Aluminium and iron fractionation of European volcanic soils by selective dissolution techniques. In: Arnalds O, Bartoli F, Buurman P, Oskarsson H, Stoops G, Garcia-Rodeja E (eds) Soils of volcanic regions in Europe. Springer, Heidelberg, pp 325–351

Gennadiev AN, Geptner AR, Zhidkin AP, Chernyanskii SS, Pikovskii YI (2007) Exothermic and endothermic soils of Iceland. Eurasian Soil Sci 40(6):595–607

Gisladottir G, Erlendsson E, Lal R, Bigham J (2010) Erosional effects on terrestrial resources over the last millennium in Reykjanes, southwest Iceland. Quatern Res 73:20–32

Gislason SR (2008) Weathering in Iceland. Jökull—Icel J Earth Sci 58:387–408

Gislason SR, Oelkers EH, Snorrason A (2006) The role of river-suspended material in the global carbon cycle. Geology 34:49–52

Gislason SR, Oekjers EH, Eiriksdottir ES, Kardjilov MI, Gisladottir G, Sigfusson B, Snorrason A, Elefsen SO, Hardardottir J, Torssander P, Oskarsson N (2009) Direct evidence of the feedback between climate and weathering. Earth Planet Sci Lett 277:213–222

Gudbergsson G (1975) Myndun móajarðvegs í Skagafirði (Soil formation in Skagafjörður, northern Iceland. J Agric Res Icel 7:20–45 (in Icelandic, English summary)

Gudbergsson G (1996) Í norðlenskri vist.Um gróður, jarðveg, búskaparlög og sögu (the influence of human habitation on soil and vegetation in three counties in North-Iceland). Icel Agric Sci 10:31–89 (in Icelandic, English abstract)

Gudmundsson Th (1978) Pedological studies of Icelandic peat soils. Unpublished PhD thesis. University of Aberdeen, Scotland

Gudmundsson Th (2009) Járnsteindir í mýrum (Iron minerals in wetlands). Fræðaþing landbúnaðarins (Agricultural Science Congress) 2009:433–437 (in Icelandic)

Guicharnaud R, Paton G (2006) An evaluation of acid deposition on cation leaching and weathering rates of an Andosol and a Cambisol. J Geochem Explor 88:279–283

Helgason B (1968) Basaltic soils of South-west Iceland II. J Soil Sci 19:127–134

Johannesson B (1960) The soils of Iceland. University Research Institute, Dept. of Agriculture reports series B, no 13. University of Iceland, Reykjavik Iceland

Johannesson B (1988) Íslenskur jarðvegur. (Icelandic Soils). Reprinted version of a book from 1960 with extended appendix. Agricultural Research Institute (now Agricultural University of Iceland), Reykjavik

Jonsdottir JF (2008) A runoff map based on numerically simulated precipitation and a projection of future runoff in Iceland. Hydrol Sci 53:100–111

Kardjilov ML, Gislason SR, Gisladottir G (2006) The effect of gross primary production, net primary production and net ecosystem exchange on the carbon fixation by chemical weathering of basalt in northeastern Iceland. J Geochem Explor 88:292–295

Larsen G, Eiriksson J (2008) Holocene tephra archives and tephrochronology in Iceland—a brief overview. Jökull—Icel J Earth Sci 58:229–250

Mangold N, Baratoux D, Arnalds O, Bardintzeff J-M, Platevoet B, Grégoire M, Pinet P (2011) Segregation of olivine grains in volcanic sands in Iceland and implications for Mars. Earth Planet Sci Lett 310:233–243

Meijer EL, Buurman P, Fraser A, Garcia-Rodeja E (2007) Extractability and FTIR-characteristics of poorly-ordered minerals in a collection of volcanic ash soils. In: Arnalds O, Bartoli F, Buurman P, Oskarsson H, Stoops G, García-Rodeja E (eds) Soils of volcanic regions in Europe. Springer, Heidelberg, pp 155–189

Monteiro F, Kleber M, Fonseca M, Madeira M, Jahn R (2007) Crystalline clay constituents of soils from European. In: Arnalds O, Bartoli F, Buurman P, Oskarsson H, Stoops G, Garcia-Rodeja E (eds) Soils of volcanic regions in Europe. Springer, Heidelberg, pp 181–195

Nanzyo M, Dahlgren R, Shoji S (1993) Chemical characteristics of volcanic ash soils. In: Shoji S, Nanzy M, Dahlgren RA (eds) Volcanic ash soils. Genesis, properties and utilization. Developments in soil science, vol 21. Elsevier, Amsterdam, pp 145–187

Oladottir BA, Larsen G, Sigmarsson O (2011) Holocene volcanic activity at Grímsvötn, Bárdarbunga and Kverkfjöll subglacial centres beneath Vatnajökull, Iceland. Bull Volcanol 73:1187–1208

Opfergelt S, Georg RB, Burton KW, Guicharnaud R, Siebert C, Gislason SR, Halliday AN (2011) Silicon isotopes in allophane as a proxy for mineral formation in volcanic soils. Appl Geochem 26: s115–s118

Oskarsson BV, Riishuus MS, Arnalds O (2012) Climate-dependent chemical weathering of volcanic soils in Iceland. Geoderma 189–190:635–651

Parfitt RL, Childs CW (1988) Estimation of forms of Fe and Al: a review, and analysis of contrasting soils by dissolution and Moessbauer methods. Aust J Soil Res 26:121–144

Rodriguez Rodriguez A, Arbelo CD, Guerra JA, Mora JL, Notario JS, Armas CM (2006) Organic carbon stock and soil erodibility in Canary Islands Andosols. Catena 66:228–235

Romans JCC, Robertson L, Dent DL (1980) The micromorphology of young soils from South-east Iceland. Geogr Ann A62:93–103

Sigbjarnarson G (1969) The loessial soil formation and soil erosion on Haukadalsheiði. Náttúrufræðingurinn 39:68–118 (in Icelandic, English summary)

Sigfusson B (2004) Assessment of *in-situ* weathering of an Histic Andosol—microcosm to field scale study. MSc thesis, Department of Geology and Geography, University of Iceland, Reykjavík

Sigfusson B, Gislason SR, Paton GI (2008) Pedogenesis and weathering rates of Histic Andosol in Iceland: field and experimental soil solution study. Geoderma 144:572–592

Sigurgeirsson MA, Arnalds O, Palsson SE, Howard BH, Gudnason K (2005) Radiocaesium fallout behaviour in volcanic soils in Iceland. J Environ Radioact 79:39–53

Simpson IA, Milek KB, Gudmundsson G (1999) A reinterpretation of the Great Pit at Hofstaðir, Iceland using sediment thin section micromorphology. Geoarchaeology 14:511–530

Stoops G, Van Driessche A (2007) Mineralogy of the sand fraction—results and problems. In: Arnalds O, Bartoli F, Buurman P, Oskarsson H, Stoops G, Garcia-Rodeja E (eds) Soils of volcanic regions in Europe. Springer, Heidelberg, pp 141–153

Stoops G, Gerard M (2007) Micromorphology. In: Arnalds O, Bartoli F, Buurman P, Oskarsson H, Stoops G, Garcia-Rodeja E (eds) Soils of Volcanic Regions in Europe. Springer, Heidelberg, pp 129–140

Stoops G, Gerard M, Arnalds O (2008) A micromorphological study of Andosol genesis in Iceland. In: Kapur S, Mermut A, Stoops G (eds) New trends in micromorphology. Springer, Heidelberg, pp 67–90

Schwertmann U, Taylor RM (1989) Minerals in soil environments. In: Dixon JB, Weed SB (eds) Soil Science Society of America book series, vol 1, 2nd edn. Soil Science Society of America, Madison, pp 379–438

Taboada T, Garcia C, Martinez-Cortizas A, Novoa JC, Potevedra X, Garcia-Rodeja E (2007) Chemical weathering of reference European volcanic soils. In: Arnalds O, Bartoli F, Buurman P, Oskarsson H, Stoops G, Garcia-Rodeja E (eds) Soils of volcanic regions in Europe. Springer, Heidelberg, pp 307–323

Thorarinsson S (1961) Wind erosion in Iceland. Ársrit Skógræktarfélags Íslands (Forestry Society Yearbook) 1960–1961. pp 17–54 (in Icelandic, Extended English Summary)

Ugolini FC, Dahlgren RA (2002) Soil development in volcanic ash. Global Environ Res 6:69–81

Van Vliet-Lanoe B, Bourgeois O, Dauteuil O (1998) Thufur formation in northern Iceland and its relation to Holocene climate change. Permafrost Periglac Process 9:347–365

Wada K, Arnalds O, Kakuto Y, Wilding LP, Hallmark CT (1992) Clay minerals of four soils formed in eolian and tephra materials in Iceland. Geoderma 52:351–365

Wolf-Boenisch D, Gislason SR, Oelkers EH, Putnis CV (2004) The dissolution rates of natural glasses as a function of their composition at 4 and 10.6, and temperatures from 25–74 °C. Geochim Cosmochim Acta 68:4843–4858

10 Frost and the Soil Environment

10.1 Arctic—Periglacial Environments

The action of frost has pronounced influence on all soil surfaces of Iceland, shaping the soils and landscapes. The Icelandic soil environment provides an overamplified example of frost-driven processes due to (i) exceptionally frequent freeze–thaw cycles and (ii) frost-susceptible soils. A variety of landforms are generated with the frost action, such as thufur (hummocks), palsas, patterned ground, and solifluction features on the surface and cryoturbation patterns within the soils. In this chapter, the nature of frost action on the soil and the geomorphic surfaces are explored, together with some of the environmental parameters that influence the frost action.

It seems that much professional effort is channeled to discussions on terminology and definitions within the domain of periglacial science. Such discussions will be limited here for purposes of simplicity, although it is important to introduce some commonly used terms. The term *periglacial* has been used in relation to areas, processes, and landforms that occur in the cold environments of the Earth. Although the term literally means 'near-glacier', this term has been broadened to include areas, landforms, and processes that are substantially affected by the action of frost; environments where frost actions and/or permafrost-related processes dominate (French 2000; Slaymaker 2011). The frost action causes displacement and mixing of soil materials. The term *cryoturbation* is often used for these processes, but the definition of the term varies considerably among literature sources (see Jones et al. 2010). While some may argue that this movement is independent of the traditional soil forming factors, it clearly depends both on climate and time and cryoturbation should be considered as part of pedogenic processes (see Bockheim et al. 2006). The result of this movement on the soil can be visible long after the climate has become much warmer—the effects of cryoturbation during the last Pleistocene glacial stage can be seen in soils where limited cryoturbation occurs today, e.g., in Europe.

A large proportion of the Earth's surface is characterized by permanently frozen soil layer, or 'permafrost', covering about 23 million km^2 (Jones et al. 2010). Soils of these areas are termed Gelisols (Soil Taxonomy) or Cryosols (WRB), defined by their permanently frozen layer (Kimble 2004). South of the permafrost areas are environments that are influenced by the action of the frost; periglacial areas affected by cryoturbation. Considering the northerly position of Iceland right under the Arctic Circle, one could assume that permafrost is a common feature of the Icelandic soil environment. Yet, permafrost is not common in Iceland except at the highest elevations. This can be thought to result from the heating effects of the oceans around Iceland, warmed up by the Gulf-Stream (see Chap. 2). However, the impacts of frost action are evident everywhere.

Another term commonly associated with northerly locations is *Arctic*. The term is derived from the Greek 'Arktos', signifying constellations of Ursa major and Ursa minor. Astronomical definition is set where the sun does not set on the summer solstice (66° 6′ N), the Arctic, or Polar Circle. However, defining the term 'Arctic' has proven to be difficult; "in biological terms the Arctic Circle is merely an abstraction" (CAFF 2001). A botanical feature often used is the tree line, which would be hard to pinpoint in reality. An average temperature of the summer month <10 °C has also been used to characterize Arctic areas (CAFF 2001), but there, one has to bear in mind that temperature fluctuations can be considerable over time. The tree line and the 10 °C line occurs at elevation somewhere above the Icelandic lowlands, the height being considerably variable between the various geographic regions, lower in the north than in the south, lower at the northern coast than inland in the north. Hence, most Icelandic environments are not far from a boundary that could be described as the Boreal—Arctic boundary (belonging to both sides), or at the boundary between the temperate and sub-Arctic regions.

The climatological characteristics of Iceland were described in Chap. 2. Its location results in a continuous 'tug of war' between the warm Atlantic Ocean air masses and the

O. Arnalds, *The Soils of Iceland*,
World Soils Book Series, DOI 10.1007/978-94-017-9621-7_10
© Springer Science+Business Media Dordrecht 2015

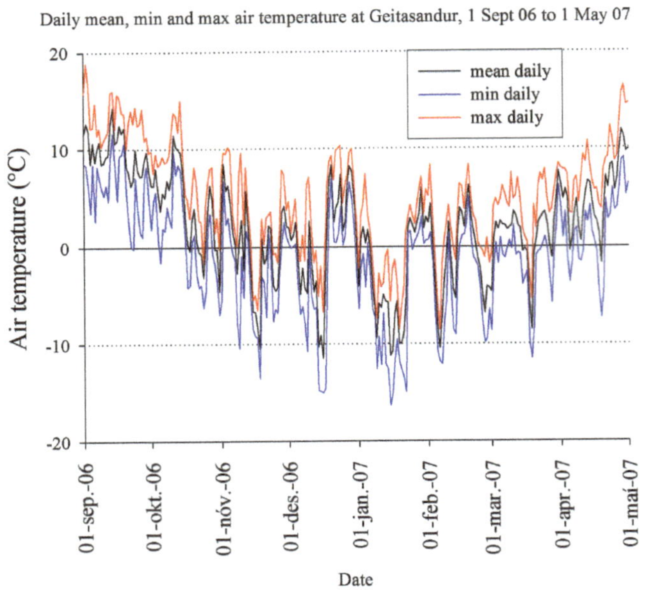

Fig. 10.1 Frequent freeze–thaw cycles over one winter at an experimental plot in South Iceland. The *lines* reflect the mean, minimum, and maximum temperatures every 24 h. Crossing the 0 °C *line* can occur more than once each day (AUI/BO/OA unpublished data)

cold Arctic air. Consequently, temperatures linger near 0 °C for extended periods during winter, especially in the lowlands. As a result, the surface is frequently freezing and thawing. It is likely that the surface of the Icelandic lowlands experience more freeze–thaw cycles each year than any other place, prompting the concept of the 'Icelandic cycle' (Washburn 1980). Winter temperatures are lower in the highlands with less frequent freeze–thaw events on the surface, also aided by more common isolative cover of snow. Figure 10.1 shows temperature fluctuations over one winter season at Geitasandur restoration experiment site (Land-Aid), near Hella in South Iceland (Arnalds et al. 2013). A comparable line for surface/soil temperatures on the continents would be warmer temperatures in summer, colder in winter, with less frequent dissection of the 0 °C line.

10.2 Water Freezes in the Soil

In order to understand why frost has such pronounced effects on Icelandic surfaces, it is helpful to look at some of the underlying principles of how water freezes in the soil. Water increases its volume by about 9 % becoming a solid matter, or ice (it is almost unique that substances become lighter changing from a liquid to a solid). This slight volume increase does not suffice in explaining the formation of thufur (hummocks; explained later) and other cryoturbation features. Water possesses many unique physical and chemical properties that are important in relation to soil frost. It contains immense amount of latent heat which is given up when the water freezes (Fig. 10.2). A token of the enormous latent heat of water is the heat transfer with the Gulf-Stream far up the North Atlantic Ocean. As the soil is cooled downward from the surface and the water begins to freeze, the latent heat is dissipated, counteracting the movement of frost downwards. This results in a temporary stationary 'freezing front' or 'freezing pane', a location where the frost

Fig. 10.2 A conceptual drawing for water freezing in soil. Latent heat is released from the water when it freezes, balancing the cold coming from above, resulting in a stationary freezing front. Water is pulled up from lower depths to the freezing front due to a water tension gradient. Considerable amount water can be drawn up to the freezing front, resulting in heaving of the soil

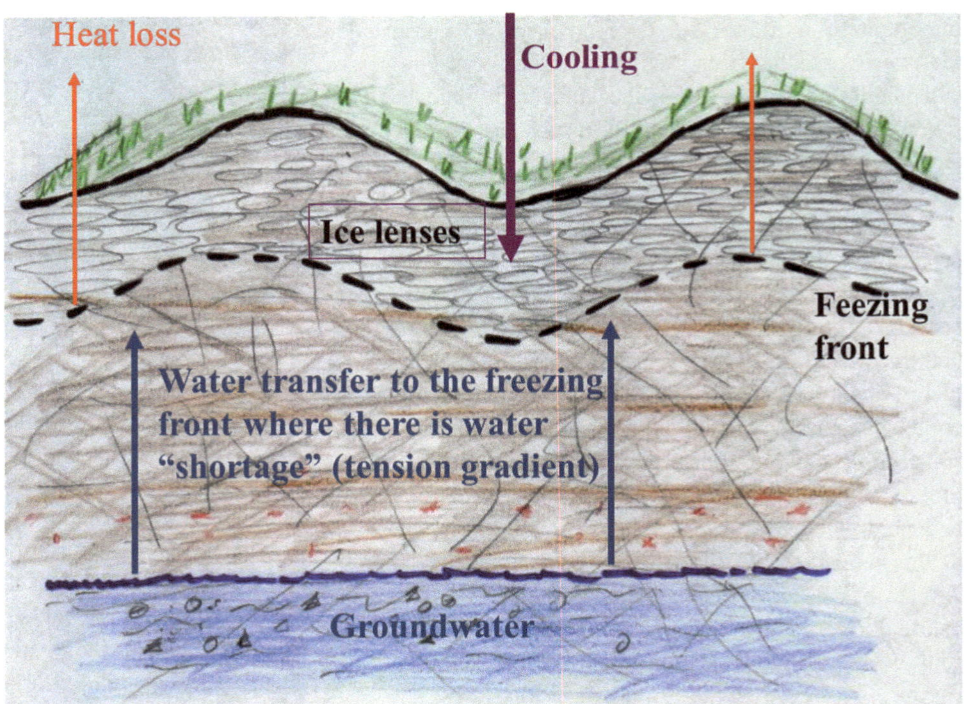

Fig. 10.3 Frost heaving. In summer, the grass surface is level with the stone-path, which has 'frost free' material under it. In winter, the grass lawn is heaved while the stone-path is not lifted except at the boundary to the grass. There is ample water in the soil under the grass in winter, enhancing the frost heave by water transfer to the freezing front

action is balanced by the energy release from the water and the coldness from above (Fig. 10.2). At this boundary, water has been immobilized by the frost, resulting in a water tension gradient (more water further down below). This leads to capillary movement of water or pull to the freezing front, sometimes referred to as 'cryosuction'. If there is ample supply of water below, a considerable amount is drawn to the freezing front and the heaving becomes exaggerated as a result. Such conditions exist where there is a short distance down to a water table (Fig. 10.3). However, if the balance between soil heat, heat released when the soil freezes, and the temperature above is changed, the freezing front may well advance still further or retreat, or even exist at more than one depth, resulting in accumulation of multiple ice-lenses in the soil.

There are other factors that contribute to the nature of cryoturbation. Thermal conductivity is important and is influenced by such factors as the organic matter content, nature of the mineral matter, water content, bulk density, vegetation and snow cover, and the temperature. Air is a poor conductor of heat. Cooling is therefore more effective in wet soils than dry, but the water can also store great amounts of heat (latent heat). Higher bulk densities usually result in faster heat transfer (less air). The great latent heat of water reduces the thermal diffusion in wet soils, slowing down the cooling effect when air temperatures become low. Vegetation that acts as windbreaks, such as shrubs and trees, accumulate snow, which provides insulation and reduces the cooling of the soils resulting in warmer soils. Organic O(H) surface horizons provide insulation that may slow spring thawing resulting in overall colder soils.

Another important factor to consider in relation to soil frost is the so-called *frost susceptibility* of the soil materials, which differ according to grain size, organic content, and composition of the materials. Coarse materials are in general not susceptible to frost as they neither store nor conduct water to the freezing front. Clay-sized materials conduct water slowly and are therefore not very frost susceptible. Silt-sized materials are the most frost-susceptible materials, having rapid hydraulic conductivities enhancing water transfer to the freezing front. Materials classified as loams, silt-loams, clay-loams, and sandy-loams are generally very susceptible to frost. The chief mineral clay constituents of Icelandic Andosols are allophane and ferrihydrite. Allophane forms stable silt-sized aggregates (Maeda et al. 1977) as is discussed in Chap. 5. Furthermore, these materials have immense water holding capacities, together with the organic matter that tends to accumulate in Andosols. The comparatively well-developed Andosols that occur under vegetation cover in Iceland are therefore very susceptible to frost. In addition, there is very high silt content in many of the desert Vitrisols (see Sect. 6.5 for Vitrisols), which also are highly frost susceptible. Both these soil types lack the cohesion provided by layer silicate clays which dominates the clay fraction in the neighboring countries, allowing for easy displacement of soil particles by the pressure from ice. Many of the soils can easily reach the liquid limit, adding to the susceptibility to frost. Many of the Vitrisols become fluid-like in the surface in the spring, when water saturates, when the silty surface soils rest on the frozen layer below, making the terrain difficult to pass, even on foot. To summarize: the soils contain lots of water, they conduct water rapidly, and lack cohesion, resulting in very frost-susceptible soils. Furthermore, environmental factors intensify the effects, such as the climatic factors. It is therefore understandable that periglacial processes are intense in Iceland.

Fig. 10.4 Ice on the ground caused by alternating freeze–thaw events over frozen ground with near zero infiltration in winter. Such ice is detrimental for plant survival in desert environments and sometimes over vegetation, causing abrasion and hindering oxygen flow to the surface. Hay fields and turf grass on soccer fields and golf courses are periodically damaged by standing ice

Fig. 10.5 Standing water on a frozen desert ground at Geitasandur LandAid research area (near Hella, South Iceland). Concrete ice with near zero infiltration rates underneath. Surface water is hardly ever seen in summer due to rapid infiltration rates. Berglind Orradottir (AUI) is sampling soil water samples for chemical analysis

Fig. 10.6 Standing water in winter in a desert near Mt. Hekla. Infiltration is impeded, thaw events cause massive runoff. No water can be seen on these surfaces of extremely fast infiltration rates in sand and lavas when the ground is not frozen. *Photo* © Elin Fjola Thorarinsdottir, ISCS

10.3 Soil Frost, Types of Ice, and Surface Runoff

Andosols under vegetation and the sandy Vitrisols of the deserts in Iceland (are Vitricryands under the Soil Taxonomy) generally have very rapid infiltration rates when not frozen, even >300 mm h^{-1} (Orradottir 2002; Orradottir et al. 2008). However, soil frost dramatically decreases the infiltration rates, to less than 50 mm h^{-1} on vegetated sites and to near zero in frozen deserts. Thus, the vegetation cover has a pronounced effect on the infiltration rates in winter, which is explained by different ice forms that form in the soil, as was shown in research conducted by Orradottir, cited above. Porous ice that is conductive to water can form in soils under rich vegetation cover. However, under less vigorous herbaceous vegetation, in depressions between thufurs and especially on deserts, concrete type of ice is formed (Orradottir et al. 2008). The reduced or blocked infiltration can result in severe runoff, which is one of the factors that shape the Icelandic desert landscapes. Flooding occurs frequently during sudden snow–thaw events where the watersheds have poor vegetation cover (Figs. 10.4, 10.5 and 10.6).

10.4 Needle-Ice Formation

The formation of needle-ice on barren surfaces is one of the most ecologically detrimental processes operating on Icelandic ecosystems. They grow vertically in or near the surface, lifting up the topmost layer, often a few millimeters thick (Fig. 10.7). In Iceland this occurs most frequently during frost-nights when there is ample moisture in the surface, such as after rainfall. An absolute prerequisite is that the surface is barren and the occurrence seems more common where the surface soil materials are relatively fine and dominated by silt, which is similar to findings elsewhere (Meentemeyer and Zippin 1981). International studies on the formation of needle-ice, which have relevance for Iceland, include those of Soons and Greenland (1970), Outcalt

Fig. 10.7 Needle-ice formations. A coin, 2.5 cm in diameter, provides a scale. The ice is layered, representing different frost events. There is ample moisture in the soil and each layer is added during frost nights

Fig. 10.8 Needle-ice formations in an erosion spot, clearly explaining the difficulty for vegetation establishment in such spots. Even small rocks are lifted from the surface

(1971), Lawler (1993), Branson et al. (1996), and Meentemeyer and Zippin (1981), with a global survey of needle-ice conditions published by Lawler (1988).

Barren conditions typical for needle-ice formation are found in erosion spots, as exemplified in Fig. 10.8 (see Chap. 12 on erosion), but they occur in a variety of other settings. Studies in Iceland on needle-ice formation are few, but include surface stability and frost-heaving measurements by Thorsson (2008), B. Orradottir, and others (AUI/ISCS, unpublished), study of the influence of frost heaving on tree nursery (Oskarsson and Brynleifsdottir 2009), and a BS project at the AUI grounds at Hvanneyri (Madsen 2013). The field experiment at Hvanneyri revealed >40 freeze–thaw events typical for needle-ice formation in just 2 months (Madsen 2013). However, water needs to be available in the surface layer of the soil, thus, lack of water (e.g., dry or frozen layer) limits ice formation, as does snow cover. Thus, needle-ice is not formed during all freezing events. They are common in spring and fall, when snow cover is limited and ample moisture is in the surface layer.

Needle-ice is capable of lifting rocks of several kilograms (Fig. 10.9), and it has a negative effect on seedlings that are struggling to get established in these surfaces. This has a detrimental influence on early ecosystem succession on barren surfaces and in part prevents natural vegetation establishment (e.g., Aradottir 1991; Oskarsson and Brynleifsdottir 2009). A biological soil crust is therefore often needed to stabilize the surface, which subsequently facilitates the development of vascular plant cover which can survive winters. An example of frost-heave measurements are presented in Fig. 10.10. Frost heave is pronounced in untreated desert plots, but fertilizer treatments have reduced frost-heave a few years after the fertilizer applications. This displacement can be >8 cm (Madsen 2013) during a single frost night. Needle-ice can grow surprisingly long as shown in Fig. 10.11, often stratified, representing consecutive frost nights. Soil particles are often incorporated into the ice.

Fig. 10.9 Needle-ice has lifted the entire surface, including pebbles and rocks. Bare Andosol surfaces are extremely susceptible to needle-ice formation

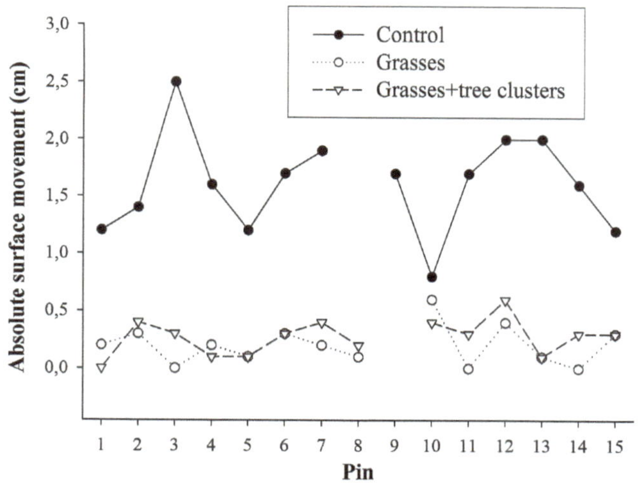

Fig. 10.10 Surface stability on barren desert and areas treated with fertilizers and grass seeds. The *Y-axis* shows vertical displacement but the *x-axis* is vertical distance with 5 cm between pins. Measurements made 7 years after treated areas were treated and fertilized. Orradottir and Arnalds, unpublished data

Needle-ice influences surface stability and vulnerability to wind and water erosion. As an example, it has been shown that after frost, threshold velocity for wind erosion was reduced from 9–10 to 6 m s^{-1} (wind at 2 m height) (Arnalds et al. 2012). Needle-ice is quite detrimental in increasing fluvial sediment loads by detaching soil particles, resulting in increased runoff from needle-ice areas (e.g., Lawler 1993), hence, the intensity of snow-melt events on frozen ground in Iceland.

10.5 Thufur

Hummocks, which in Icelandic are termed *thufur* (spelled 'þúfur' in Icelandic, single 'þúfa') are prominent and often dominating features of the Icelandic soil surface. Thufur are mounds or hummocks, often 50–150 cm in diameter. They can be low (<10 cm) to more than 1 m height, of various steepness (Fig. 10.12). Examples of thufur interior is shown in Fig. 10.13. Thufur are found on the surface of nearly all vegetated surfaces in Iceland, being formed in the aeolian-andic soil mantle (volcanic loess) that rests on other types of bedrock. The soils include Brown, Gleyic, and Histic Andosols, but also on Histosols to some degree. Thufur occurs in Iceland under a range of average annual temperatures from >5 to <−2 °C.

Harris et al. (2009) defined hummocks as "Dome-shaped features with raised center and depression or through between hummocks," but they note that the terminology used for patterned ground is confusing. Schunke and Zoltai (1988) suggested that the term 'earth hummocks' were hummocks of permafrost areas while 'thufur' were hummocks in areas with seasonal frost. The Icelandic term 'thufur' has increasingly become the international term for the types of hummocks that occur in Iceland, and have been used in research describing these features on the slopes of a volcano in Korea (Kim 2008) and in South Africa (Grab 2005) to give examples of the international use of the term. Hummocks are also called 'pounus' in Fennoscandinavia

Fig. 10.11 Exceptionally long needle-ice. Soil and a plant on the top! *Photo* Axel Arnason Njardvik, © Sigthrudur Jonsdottir

Fig. 10.12 Example of thufur surfaces in a poor heathland vegetation. Thufur are among the most common and influential features of Icelandic soil surfaces. Photo: Fanney Osk Gisladottir/ AUI

(e.g., Luoto and Seppäla 2002). Van Vliet-Lanoë et al. (1998) considered thufur as a subtype of hummocks, but they fall well within a definition suggested by Grab (2005) as miniature cryogenic mounds generally less than 1.5 m in height, which form both in seasonally frozen and permafrost areas. Definitions related to thufur and hummocks will undoubtedly still be the focus of continuing discussions in the literature.

The formation of thufur is dependent on ample soil water supply and soil frost. The northern boundary seems to be related to the northern vegetation boundary (Schunke 1977). An annual temperature lower than 3–6 °C has been suggested as a prerequisite for thufur formation (Grab 2005), but lower values were suggested by Tarnocai and Zoltai (1978). Icelandic annual temperatures fall well within the temperatures suggested in these papers.

Thufur in Iceland were noted early on by scientists, such as Gruner (1912) and Thoroddsen (1913), and they occur in earlier nonscientific writings, while their nature was poorly understood early on. Schunke (1977) made a detailed study of the morphology of thufur in Iceland, but otherwise, studies related to thufur formation in Iceland are relatively few.

10.5.1 Thufur over Shallow Water Table

The conceptual model provided in Sect. 10.2 shows that thufur are likely to form where standing water is within reach of capillary pull to the freezing front. This was noted early by Johannesson (1960), who suggested that the tallest thufur occur at the transition zone between wetlands and heathlands, which is quite evident on short catenas from wetlands in depressions to dry hilltops. Figure 10.14 shows such transect measured from wetland surface to heathland above it, with the highest thufur closest to the wetland. It is noteworthy that thufur do not form or are at least not substantial within the wetland areas, where there is standing water or the water table is within a few centimeters from the surface. The thufur appear to be highest were the groundwater table is at optimal depth, somewhere between 20 and 60 cm. This fits well with the Johannesson (1960) model.

10.5.2 'Dryland Thufur'—Thufur in Areas Without the Presence of Shallow Water Table

In the discussion above, the role of groundwater table was emphasized, a model that is well established. However, only a part of the soil in Iceland has water table within the reach of capillary pull. However, thufur occur nearly everywhere; there are extensive areas where thufur occur without the water table being close to the surface (Brown Andosols; Fig. 10.15). This calls for alternative explanations for the formation of thufur under such drylandconditions, why they are so common, and why they can become as large as >1 m in height and diameter under such conditions. An example of

Fig. 10.13 A soil profile exhibiting an interior of a thufur

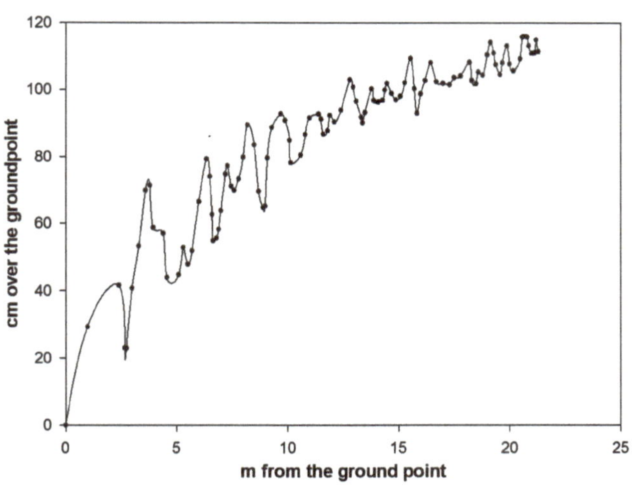

Fig. 10.14 A transect running from open wetland to dry heathland at Mosfellsheiði, West Iceland. The thufur are highest in the transition zone near the wetland zone. From Arnalds and Sigurjonsdottir (2012)

such thufur are in soils of the Holocene lavas between the Krafla area and Kelduhverfi in Northeast Iceland, where the water table is often at 20–70 m depth, yet there some of the largest thufur in Iceland are found. Thufur in this area were studied by Van Vliet-Lanoë et al. (1998).

An important factor in explaining the formation of thufur without the presence of shallow water table is high water content in the soil for movement to the freezing front, but also different conditions prevailing under each mound compared to the space between them, resulting in 'differential frost heaving' (see Grab 2005). Walker et al. (2008) emphasized the difference in vegetation and snow cover characteristics from the 'on-top' to the 'between thufa' locations, which affects the thermal and water regimes in the soils, resulting in water being drawn to the 'center of the patterns'. This would apply both to the 'water table thufur' and the 'dryland thufur'. Below are listed some important characteristics, properties, and processes that contribute to the formation of thufur without the presence of shallow water table based on review in Icelandic by Arnalds (2010a).

(i) The soils are Andosols, with extremely high water retention, but lacking cohesion, allowing for easy movement or squeeze of the soil materials, especially when near saturation (when liquid limit is easily reached).

(ii) The clay materials are dominated by allophane, which forms stable silt-sized aggregates, with high hydraulic conductivity, yet also high water retention. The result is very frost-susceptible materials. Coarse tephra layers, where present, do halt the capillary movement toward the freezing front (lower thufurs within the active volcanic belt).

(iii) Extremely common freeze–thaw cycles enhance water accumulation in the soil all winter long; water is both transferred from the soils below up to the frozen area, and added to the top during thaw events. The thaw events are often characterized by both rainfall and snow–thaw, adding water to the soils. Yet the soil has a frozen layer, which slows downward movement of the winter precipitation to lower layers, resulting in water saturated surface above the frozen layer.

(iv) Animal grazing is clearly a factor adding to the formation of thufur in Iceland (both water table and dryland soil conditions). Domestic animals, especially the heavy ones such as horses and cattle, but also sheep to some degree, nearly always put their feet down between the thufur. As the soil has low cohesion, there is a gradual push upwards of the thufur around the depression. The highest and steepest thufur are found within areas heavily and often overgrazed by horses; grazing often takes place in winter when the soil is relatively saturated by water. Horse grazing therefore

Fig. 10.15 Example of thufur in area where the water table is far below the surface. The *photo* is from Kelduhverfi, NE Iceland, and the water table is several tens of meters below the surface. Yet, this area has very large thufurs

accelerates the process (Figs. 10.16 and 10.17). It is interesting to note that thufur occur in the Azores Islands where the soil does not freeze, but the soils are Andosols, often of extreme water holding capacities (Hydric Andosols) with low cohesion—and are grazed by heavy cattle (see Arnalds 2010a). The influence of grazing animals is undoubtedly an underestimated factor in the international literature concerning thufur development.

After thufur development has begun, there are increasingly different conditions in the soils between the thufur than within them. There is shelter in the depressions which also accumulate snow, but the thuufur are more exposed to the cold, allowing for earlier formation of a freezing front. Vegetation cover often provides less insulation on top of thufur than in-between areas. However, there is high probability of periodic formation of standing water between the thufur during thaw events (Orradottir et al. 2008).

Fig. 10.16 A fence-line between an area heavily grazed by horses and a moderately grazed area by sheep. Thufur have formed within the horse pasture as a result of hoof action, vegetation removal (grazing), climatic factors (modified by the grazing), and soil properties

Fig. 10.17 Steep, recent thufur (10–20 years) in a heavily grazed horse pasture

10.5.3 Thufur, Geography, and Some General Considerations

Based on this model, larger hummocks would be expected where the soils are relatively fine textured with more allophane clay constituents, which is extremely frost-susceptible material. In areas with dryland thufur surfaces (without shallow water table), thufur are larger at a distance from aeolian sources and where soil development has advanced further compared to areas experiencing rapid aeolian accumulation and coarse tephra fallout. The northeastern, north, and western areas meet these criteria in many places, with thufur dominating the landscape regardless of the depth to the water table. The coarse tephra layers deposited near the most active volcanoes such as Hekla and Katla seem to reduce thufur formation. Thufur are not as distinct in dryland soils of the southern communal grazing areas, such as east of Mýrdalsjökull glacier where relatively level heath and grassland surfaces are common, even with water table close to the surface. Deep isolative snow cover would reduce the formation of thufur as noted by Thorarinsson (1951), and the author has noticed that thufur do not form or are much lower inside birch forests than outside within the same area, but the forest both shelters from wind and accumulates more snow in winter, but they are also not grazed by heavy animals.

Thufur make the landscape challenging to walk in, and they made hay-making difficult. Yet, farmers in the old days were reluctant to level out the thufur surfaces to make hay-making easier, as many believed that the increased surface area would provide more yields than level land. Leveling of thufur surfaces with machinery was among the first major projects of the new age in agriculture in Iceland. Some areas had particularly notorious thufur; it was said about one farm in Northeast Iceland that the thufur were so steep that even cats broke their legs in the terrain.

The stratigraphy within the soils in Iceland, created by steady aeolian and periodic tephra inputs, is often quite clear in the soil strata. In many lowland areas, the author has observed that tephra layers are often relatively straight horizontal in strata dating up to the Middle Ages, but become cryoturbated in the surface horizons, but the depth for non-cryoturbated soil horizons is often greater in the highlands, e.g., >2,500 years, as was noted by van Vliet-Lanoë et al. (1998). This may both be indicative of land use and vegetation cover changes (such as removal of birch forests and increased grazing) and climate change with cooler climate during the Middle Ages.

10.6 Solifluction

Solifluction is the slow, downward movement of the soil surface due to the action of freeze–thaw processes and the pull of gravity. Several landforms, such as terraces ('stallar' or 'paldrar' in Icelandic; Fig. 10.18) and lobes ('jarðsilstungur'; Fig. 10.19) are formed. These features are prominent on nearly all vegetated slopes in Iceland (Figs. 10.20 and 10.21) and barren slopes, where texture is rather fine and

Fig. 10.18 Example of solifluction terraces in Iceland. One way to look at the terraces is to view them as wave motion of the soil materials down the slope (gravitational pull) in a similar fashion as small ocean waves (wind push)

Fig. 10.19 Example of solifluction lobes in Iceland. The front of the lobes often has turf and gravel that impedes the downward movement, causing the enlargement of the lobes. The lobes are very susceptible to landslides upon disturbance when water saturated

screen movement does not mask the effects of solifluction. Lobes commonly have turf or coarse materials at the forefront of the lobes providing cohesion against the downward pull, so-called stone-banked and turf-banked lobes. Downward movement has been reported as few as >60 mm year^{-1} in the literature outside of Iceland (e.g., Benedict 1970, 1976; Ridefelt et al. 2009), but rates have not been measured in Iceland. Needle-ice formation is considered to be an important factor for solifluction movement, at least of the barren surfaces (see general review by Matsuoka 2011).

When soilerosion was mapped in all of Iceland (Arnalds et al. 2001), erosion spots associated with solifluction was a special entity of the erosion classification, because barren soil on solifluction slopes are exposed to running water and are more unstable than on level land.

As with the thufur formation, one has to keep the special characteristics of Icelandic Andosols in mind, the highly frost-susceptible material, high water holding capacity, and lack of cohesion when considering solifluction processes. These factors, associated with ideal temperature and soil

Fig. 10.20 A combination of thufur and solifluction terraces, created under frost-heave in the influence of the gravity of the slope. The thufur forms are >1 m high

moisture conditions, are likely to add to the activity of solifluction in Iceland. Barren ground can develop on top of the terraces, but the turf at the forefront still provides cohesion. The slopes can therefore look well-vegetated looking up the slopes, but only half-vegetated when looking down the slopes. This is quite noticeable in Northwest Iceland. Furthermore, solifluction slopes are quite susceptible to slope failures, with evidence of landslide being a common feature in areas characterized by solifluction lobes.

10.7 Patterned Desert Ground

Soil frost has a way to create peculiar patterns in the surface of cold soils, including 'patterned ground', 'sorted polygons' and ice wedges. Patterned ground "is a group term for the more or less symmetrical forms, such as circles, polygons, nets, steps, and satrapies, that are characteristic of, but not necessarily confined to, mantle subject to intensive frost action" (Washburn 1956). These geomorphic phenomena are among the distinctive surface features of the Arctic areas and are explained in most general textbooks on geomorphology and texts on cryology (see Ritter et al. 1996; Goldthwait 1976). Polygons have rocky gutters or boundaries with finer-textured centers (e.g., Goldthwait 1976). Polygons form gradually with differential action of soil frost on finer versus more coarse textured soil materials. When glacial till appears from underneath the retreating glacier, or when erosion removes the aeolian–andic soil profile from the underlying surface (see Chap. 12 on soil erosion), soil frost begins to push gradually the more coarse materials to the sides, while finer materials are left or accumulate (relatively speaking) at the center. The finer materials, mainly silt with some allophane clay, hold more soil moisture than the edges, resulting in more and more active differentiation process. The process is intensified by the extreme frost susceptibility of the silty materials at the center—and most likely the unique frequency of freeze–thaw cycles, adding ever more moisture to the system in Iceland. Figure 10.22 shows newly formed small patterns. The picture is taken mid-morning during a dry day after a cold night, with the finer materials in the center still retaining moisture while the gravelly edges have dried up. The center area is sometimes called 'frost boils' or 'mud boils' (Fig. 10.23), being extremely unstable and creating conditions similar to quicksand when the boils are water saturated over frozen layer, such as in the spring.

It appears from my travels in Iceland that sorted polygons are much more noticeable in areas receiving relatively low amount of aeolian sedimentation (lag-gravel, melur), while they are often absent or at least less prominent in the more sandy areas of the volcanic belt (sandy lag-gravel, sandmelur and other sandy surfaces). The polygons also rely on deep enough regolith ('loose materials') of >20–30 cm (Priesnitz and Schunke 1983). It is not clear if there are any climatic or regional restrictions to the distribution of sorted polygons in Iceland. An elevation restriction of >200–300 m has been suggested (Thorarinsson 1964; Friedman et al. 1971), but it is clear that polygons occur at sea level in most parts of the

Fig. 10.21 A mountain slope characterized by solifluction terraces

Fig. 10.22 Small newly formed patterned ground in the western lowlands after erosion removed an Andosol mantle and exposed underlying gravelly material. *Photo* taken in the morning, with moisture retained in the finer-textured center of the polygons, revealing the textural differences. *Photo* Fanney Osk Gisladottir, AUI

Fig. 10.23 Typical polygons (patterned ground) at about 200 m elevation in West Iceland. A *black lid* from a camera lens provides the scale. The center of the polygons is frost-susceptible silty materials, with considerable cryoturbation occurring during the winter, pushing the more coarse materials aside

country on barren ground where sand accumulation is limited. However, Thorarinsson (1964) concluded that large-scale subsoil polygons found on low-level coastal plains and valley bottoms are not formed under current conditions, but are rather 'fossils' from colder periods. This can, however, be doubted, noting the examples in Fig. 10.22, where polygons are forming on surfaces that have recently become barren at a lowland location in the west. One can conclude that further research is needed to pinpoint the distribution and the conditions for the formation of patterned ground on deserts in Iceland. Again, it is likely that high moisture content throughout winter, frequent freeze–thaw cycle with the occurrence of periods of low temperatures (<−10 °C), and frost-susceptible materials are factors that need to be considered.

Frost heaving has an important bearing for many of the desert surfaces of Iceland, as it maintains the gravelly to rocky surfaces. In many of the desert areas there is a rapid influx of aeolian materials from various dust sources (Arnalds 2010b) which are added on top of the desert surfaces, often 0.01 to >1 mm year^{-1}. Frost action pushes the gravel and rock fragments continuously to the surface each year, thus maintaining the gravelly surface. As a result, the surface is rising, creating sandy subsurface horizons under a gravelly surface layer. However, some of these surfaces become unstable during the most intense dry storms, and rapid snowmelt events, counteracting the sand accumulation.

Ice wedges, which form frost cracks in the ground (see Goldthwait 1976) do occur in Iceland (Thorarinsson 1964; Friedman et al. 1971), but they are not as widespread as the polygons.

10.8 Palsas

Palsas are ice cored mounds, >50 cm, that rise up from wetlands in frost affected areas (Seppälä 1988). Some definitions restrict palsas to soils having peat horizons (Seppälä 1988), which is not always the case in Iceland (see below). Palsas are common in the circumpolar regions and they have received considerable research attention (e.g., Seppälä 1988; Pissart 2002; Luoto et al. 2004; Zuidhoff and Kolstrup 2005). Palsas are remarkable geomorphic features, but they are now subjected to degradation in many areas of the world because of warmer climate brought on by climate change

Fig. 10.24 The Orravatn palsa area in the highlands north of Hofsjökull glacier

Fig. 10.25 A possible inactive rock glacier, with activity during colder periods. The nature of many such features in Iceland are disputed (either rock glaciers or landslides/rock falls). Photo: Asa L. Aradottir, AUI

(e.g., Vallée and Payette 2007; Luoto et al. 2004). There are numerous palsa areas in the Icelandic highlands, but the best studied are the prominent Þjórsárver palsas south of Hofsjökull glacier (Thorhallsdottir 1994, 1996, 1997), but some of the largest palsas are found in the Orravatnsrústir area north of Hofsjökull glacier (Saemundsson et al. 2012). Figure 10.24 shows palsas in the Orravatnsrústir area. Palsas occur sporadically in the highlands of the northern part of Iceland such as in the Blanda area (North), within vegetated patches on the Jökuldalsheiði and on Fljótsdalsheiði (East Iceland), northeast of the Vatnajökull glacier. The distribution of palsas in Iceland is declining; palsas have decreased or disappeared from certain areas over the past 5–50 years (Bergmann 1973; Arnalds 2010a). Palsa areas are important ecosystems in Iceland in spite of limited distribution because of their scientific (including climate change evidence) and aesthetic values, rarity, and diverse wildlife (Magnusson et al. 2009).

The Icelandic palsas form where there is an isolated cover of vegetation at highland locations and where there is ample water near the surface (see Thorhallsdottir 1996; Saemundsson et al. 2012). Most of these palsas form in areas receiving a large amount of aeolian dust input in addition to periodic tephra deposition events, which separates them from many other palsa areas of the world (Saemundsson et al. 2012). This results in soils that have lower organic content (often <5 % C in surface horizons) compared to the organic Arctic soils where palsas are common elsewhere. However, the presence of peat is commonly used in definitions (e.g., Pissart 2002) or mentioned as a requirement for palsa formation (e.g., Zuidhoff and Kolstrup 2005). Saemundsson et al. (2012) stated that "the surface root mat of the wetland vegetation patches in the Icelandic highlands seems to have similar thermal and hydraulic properties as peat in the Arctic." Discontinuous snow-cover with low summer temperatures also contributes to the formation of palsas in Iceland.

Palsa areas are known to exhibit cyclic behavior with alternating growth and decay (e.g., Seppälä 1986, 1988; Zuidhoff and Kolstrup 2000). Thorhallsdottir (1996) and Kristinsson and Sigurdardottir (2002) emphasized the dynamic nature of the palsas in the Þjórsárver region (south of Hofsjökull glacier), with the soils becoming dryer as the ground rises with the formation of the palsa. Draining of the soil surface causes vegetation changes that subsequently reduce insulation, hence, the wet climate, leading to melt of the palsa and ultimately a pond is formed.

Many Icelandic palsa areas are under threat of hydropower development, such as the areas in Þjórsárver and Orravatn.

10.9 The Rock Glacier Dilemma

Features that resemble rock glaciers are very common in Iceland, especially in North Iceland, but also elsewhere in mountainous landscapes, such as in the Westfjords, West, Northeast, and East Iceland. These features are traditionally explained as rockslides or rock falls as described in the works of Olafur Jonsson (see Jonsson 1976). Some features, such as the gigantic Vatnsdalshólar (North Iceland; Jonsson et al. 2004) and the massive gravel area in Loðmundarfjörður in East Iceland (dated 1,000–2,000 years old; Guttormsson 2008; Hjartarson 1997) have been identified as rock falls. More recently, many of these features have been considered as classical rock glaciers and active rock glaciers are found at high elevations in the central north (Gudmundsson 1995, 2005). However, there is considerable disagreement among earth scientists about many of the features, which are either considered as rockfalls/rock slides or rock glaciers (often old and inactive). An example of astounding natural features that both have been considered as a rockslide and a rock glacier is the 'Stóraurð' under the Dyrfjöll mountains (Door-mountains) in East Iceland (Fig. 10.25).

Fig. 10.26 It is necessary to place frost-free materials below concrete and buildings. Failure to use proper materials causes frost damage like seen on this asphalt path

10.10 Permafrost

There has been limited research concerning the extent of pSermafrost in Iceland except the previously research on palsa areas. The most comprehensive account of permafrost in Iceland was published by Farbrot et al. (2007), which was based on short time series from boreholes at high altitudes. Their results show that permafrost is widespread at elevations above 900 m in northern and eastern Iceland. They concluded that permafrost in Iceland was quite sensitive to global warming.

10.11 Construction and Soil Frost

Considering the frost susceptibility of soils in Iceland, it is not surprising that they have to be removed from construction sites, all the way down to solid bedrock or some other frost-free materials. The soils are replaced by sand and gravel before construction. This adds considerable costs to construction in Iceland, such as for roads and buildings (Fig. 10.26).

References

Regarding punctuation and Icelandic characters in citations: See note on punctuation in the Preface

Aradottir AL (1991) Population biology and stand development of birch (*Betula pubescens* Ehrh.) on disturbed sites in Iceland. Unpublished PhD dissertation, Texas A&M University, College Station, Texas

Arnalds O (2010a) Cryoturbation, frost and soil (Kulferli, frost og mold). Agricultural University of Iceland Publication no 26 (in Icelandic). http://www.lbhi.is/sites/default/files/gogn/vidhengi/thjonusta/utgefid_efni/RitLbhi/Rit_LBHI_nr_26.pdf

Arnalds O (2010b) Dust sources and deposition of aeolian materials in Iceland. IcelAgric Sci 23:3–21

Arnalds O, Sigurjonsdottir M (2012) Þúfurnar á Mosfellsheiði (The Mosfellsheidi thufurs). Agricultural University of Iceland publication no 39 (in Icelandic). http://www.lbhi.is/sites/default/files/gogn/vidhengi/thjonusta/utgefid_efni/RitLbhi/rit%20lbhi%20nr%2039.pdf

Arnalds O, Thorarinsdottir EF, Metusalemsson S, Jonsson A, Gretarsson E, Arnason A (2001) Soil erosion in Iceland. The Soil Conservation Service (Landgræðsla ríkisins) and the Agricultural Research Institute (Rala), Reykjavík. http://jardvegsstofa.lbhi.is/lisalib/getfile.aspx?itemid=1807 (Translation of the original 1997 publication)

Arnalds O, Gisladottir FO, Orradottir B (2012) Determination of aeolian transport rates of volcanic soils in Iceland. Geomorphology 167–168:4–12

Arnalds O, Orradottir B, Aradottir A (2013) Carbon accumulation in Icelandic desert Andosols during early stages of restoration. Geoderma 193–194:172–179

Benedict JB (1970) Downslope soil movement in a Colorado Alpine Region: rates processes, and climate significance. Arct Alp Res 2:165–226

Benedict JB (1976) Frost creep and gelifluction features: a review. Quat Res 6:55–76

Bergmann B (1973) Um rústir á húnvetnskum heiðum (Notes on palsas on the North-west highlands). Náttúrufræðingurinn 42:190–198 (in Icelandic)

Bockheim JG, Mazhitova G, Kimble JM, Tarnocai C (2006) Controversies on the genesis and classification of permafrost-affected soils. Geoderma 137:33–39

Branson J, Lawler DM, Glen JW (1996) Sediment inclusion events during needle ice growth: A laboratory investigation of the role of soil moisture and temperature fluctuations. Water Resour Res 32:459–466

CAFF (Conservation of Arctic Flora and Fauna) (2001) Arctic flora and fauna: status and conservation. Edita, Helsinki

Farbrot H, Etzelmüller B, Schuler TV, Gudmundsson A, Eiken T, Humlum O, Björnsson H (2007) Thermal characteristics and impact of climate change on mountain permafrost in Iceland. J Geophys Res 112:F03S90. doi:10.1029/2006JF000541

French HM (2000) Does Lozinski's periglacial realm exist today? A discussion relevant to the usage of the term "periglacial". Permafrost Periglac Process 11:35–42

Friedman JD, Johansson CE, Oskarsson H, Svensson H, Thorarinsson S, Williams RS Jr (1971) Observations on Icelandic polygon surfaces and palsa areas. Photo interpretation and field studies. Geografiska Annaler Ser A Phys Geogr 53(3/4):115–145

Goldthwait RP (1976) Frost sorted patterned ground: a review. Quat Res 6:27–35

Grab S (2005) Aspects of the geomorphology, genesis and environmental significance of earth hummocks (thúfur, pounus): miniature cryogenic mounds. Prog Phys Geogr 29:139–155

Gruner PM (1912) Die Bodenkultur Islands. Archiv für Biontologie, Gesellschaft Naturforschender Freunde Zu Berlin, III. Band, 2. Heft. In Kommission Bei R. Friedländer & Sohn, Berlin

Gudmundsson A (1995) Berghlaup eða urðarjöklar? Rock falls or rock glaciers? Náttúrufræðingurinn 64:177–186 (in Icelandic)

Gudmundsson A (2005) Dreifing þykkra urðarbingja í fjalllendi á Íslandi. (Thick gravel-rock formations in Icelandic mountains) Vorráðstefna Jarðfræðifélags Íslands, ágrip erinda og veggspjalda, pp 10–11 (in Icelandic)

Guttormsson H (2008) Úthérað, ásamt Borgarfirði eystra, Víkum og Loðmundarfirði. Árbók Ferðafélags Íslands 2008 (the Icelandic touring association yearbook 2008). ISSN 0256-8470

Harris C, Arenson LU, Christiansen HH, Etzelmuller B, Frauenfelder R, Gruber S, Haeberli W et al (2009) Permafrost and climate in Europe: Monitoring and modelling thermal, geomorphological and geotechnical responses. Earth Sci Rev 92:117–171

Hjartarson A (1997) Loðmundarskriður. The Loðmundarfjörður rock slide, East Iceland. Náttúrufræðingurinn 67:97–103 (English summary)

Johannesson B (1960) Soils of Iceland. University Research Institute, Dept. of Agriculture, reports series B, no 13

Jones A, Stolbovoy V, Tarnocai C, Broli G, Spaargaren O, Montanarella L (eds) (2010) Soil atlas of the northern circumpolar region. European Commission, Publications Office of the European Union, Luxembourg

Jonsson O (1976) Berghlaup (Rockfalls). Ræktunarfélag Norðurlands, Akureyri (in Icelandic)

Jonsson HB, Nordahl H, Petursson HG (2004) Myndaði berghlaup Vatnsdalshóla? (Was it a Rock Avalanche that formed Vatnsdalshólar in Northern Iceland?) Náttúrufræðingurinn 72:129–138 (In Icelandic, English Summary)

Kim T (2008) Thufur and turf exfoliation in a subalpine grassland on Mt Halla, Jeju Island, Korea. Mt Res Dev 28:272–278

Kimble JM (2004) Preface. In: Kimble JM (ed) Cryosols. Permafrost-affected soils. Springer, Berlin

Kristinsson H, Sigurdardottir R (2002) Freðmýrarústir á áhrifasvæði Norðlingaölduveitu. Breytingar á 30 ára tímabili. (Palsas near the Norðlingaalda water diversion. 30 yr changes). The Icelandic Institute for Natural History, NÍ-02002, Reykjavík, Iceland (in Icelandic)

Lawler DM (1988) Environmental limits of needle-ice: a global survey. Arct Alp Res 20:137–159

Lawler DM (1993) Needle ice processes and sediment mobilization on river banks: the River Ilston, West Glamorgan, UK. J Hydrol 150:81–114

Luoto M, Seppälä M (2002) Characteristics of earth hummocks (pounus) with and without permafrost in Finnish Lapland. Geogr Ann 84A:127–136

Luoto M, Fronzek S, Zuidhoff F (2004) Spatial modeling of palsa mires in relation to climate in northern Europe. Earth Surf Proc Land 29:1373–1387

Madsen KS (2013) Needle ice in Icelandic Andosols: a field experiment in Hvanneyri, Iceland. Unpublished BS thesis, University of Copenhagen and the Agricultural University of Iceland

Maeda T, Takenaka H, Warkentin BP (1977) Physical properties of allophane soils. Adv Agron 29:229–264

Magnusson SH, Magnusson B, Olafsson E, Gudjonsson G, Gudmundsson GA, Kristinsson ., Egilsson K, Skarphedinsson KH, Heidmarsson S, Ottosson JG (2009) Vistgerðir á miðhálendi Íslands. Flokkun, lýsing og verndargildi. (Ecotypes in the Central Highlands of Iceland. Classification, description and conservation value). Icelandic Institute of Natural History NÍ-09008, Reykjavík, Iceland (in Icelandic)

Matsuoka N (2011) Climate and material controls on periglacial soil processes: toward improving periglacial climate indicators. Quat Res 75:356–365

Meentemeyer V, Zippin J (1981) Soil moisture and texture controls of selected parameters of needle ice growth. Earth Surf Process Landf 6:113–125

Orradottir B (2002) The influence of vegetation on frost dynamics, infiltration rate and surface stability in Icelandic rangelands. MSc thesis, Texas A&M University, College Station, Texas

Orradottir B, Archer SR, Arnalds O, Wilding LP, Thurow TL (2008) Infiltration in Icelandic Andisols: the role of vegetation and soil frost. Arct Antarct Alp Res 40:412–421

Oskarsson H, Brynleifsdottir SJ (2009) The interaction of fertilization in nursery and field on survival, growth and the frost heaving of birch and spruce. Icel Agric Sci 22:59–68

Outcalt SI (1971) An algorithm for needle ice growth. Water Resour Res 7:394–400

Pissart A (2002) Palsas, lithalsas and remnants of these periglacial mounds. A progress report. Prog Phys Geogr 26:605–621

Priesnitz K, Schunke E (1983) The significance of periglacial phenomena in Iceland. Polarforschung 53:9–19

Ridefelt H, Åkerman J, Beylich AA, Boelhouwers J, Kolstrup E, Nyberg R (2009) 56 years of solifluction measurements in the Abisko Mountains, northern Sweden—analysis of temporal and spatial variations of slow soil surface movement. Geografiska Annaler 91 A:215–232

Ritter DF, Kochel RC, Miller JR (1996) Process geomorphology, 3rd edn. WCB Publishers, Boston

Saemundsson Th, Arnalds O, Kneisel C, Jonsson HP, Decaulne A (2012) The Orravatnsrustir palsa site in Central Iceland—palsas in an aeolian sedimentation environment. Geomorphology 167–168:13–20

Schunke E (1977) Zur Genese der Thufur Islands und Ost-Grönlands. Erdkunde 31:297–287
Schunke E, Zoltai SC (1988) Earth hummocks (thufur). In: Clark MJ (ed) Advances in periglacial geomorphology. John Wiley, New York, pp 231–245
Seppälä M (1986) The origin of palsas. Geografiska Annaler Ser A 68:141–147
Seppälä M (1988) Palsas and related forms. In: Clark MJ (ed) Advances in periglacial geomorphology. John Wiley, New York, pp 247–297
Slaymaker O (2011) Criteria to distinguish between periglacial, proglacial and paraglacial environments. Quaestiones Gographicae 30:85–94
Soons JM, Greenland DE (1970) Observations on the growth of needle ice. Water Resour Res 6:579–593
Tarnocai C, Zoltai SC (1978) Earth hummocks of the Canadian Arctic and Subarctic. Arct Alp Res 10:581–594
Thorarinsson S (1951) Notes on patterned ground in Iceland, with particular reference to the Icelandic 'flás'. Geogr Ann 33:144–156
Thorarinsson S (1964) Additional notes on patterned ground in Iceland with a particular reference to ice-wedge polygons. Biul Peryglac 14:327–333
Thorhallsdottir ThE (1994) Effects of changes in groundwater level on palsas in Central Iceland. Geografiska Annaler Ser A Phys Geogr 76 (3):161–167
Thorhallsdottir ThE (1996) Seasonal and annual dynamics of frozen ground in the central highland of Iceland. Arct Alp Res 28:237–243
Thorhallsdottir ThE (1997) Tundra ecosystems of Iceland. In: Wiegolaski FE et al (eds) Polar and Alpine tundra. Elsevier, Amsterdam, pp 85–96
Thoroddsen Th (1913) Polygon boden und 'thufur' auf Island. Petermans Geographische Mitteilungen 59:253–255
Thorsson J (2008) Desertification of high latitude ecosystems: conceptual models, time-series analyses and experiments. PhD dissertation, Rangeland Ecology and Management, Texas A&M University, College Station, Texas (Unpublished)
Vallée S, Payette S (2007) Collapse of permafrost mounds along a subarctic river over the last 100 years (northern Québec). Geomorphology 90:162–170
van Vliet-Lanoë B, Bourgeois O, Dauteuil O (1998) Thufur formation in northern Iceland and its relation to Holocene climate change. Permafrost Periglac Process 9:247–365
Walker DA, Epstein HE, Romanovsky VE, Ping CL, Michaelson GJ, Daanen RP, Shur Y, Peterson RA, 13 other authors (2008) Arctic patterned-ground ecosystems: a synthesis of field studies and models along a North American Arctic Transect. J Geophys Res 113:G03S01:1–17. doi:10.1029/2007JG000504
Washburn AL (1956) Classification of patterned ground and review of suggested origins. Bull Geol Soc Am 67:823–866
Washburn AL (1980) Geocryology. John Wiley, New York
Zuidhoff F, Kolstrup E (2000) Changes in palsa distribution in relation to climate change in Laivadalen, northern Sweden, especially 1960–1997. Permafrost Periglac Process 11:55–69
Zuidhoff F, Kolstrup E (2005) Palsa development and associated vegetation in Northern Sweden. Arct Antarct Alp Res 37:49–60

The Volcanic Aeolian Environments of Iceland 11

11.1 Introduction

Nowhere outside of the arid regions on Earth is nature as much influenced by wind erosion processes as in Iceland. A large proportion of Icelandic surfaces are made of unstable sandy deposits (Fig. 11.1), and dust is continuously being redistributed over the entire country, having a dominant influence on soils and ecosystems. Thus, to understand the soils and Icelandic nature in general, one has to give the sandy desert a special consideration. In soil science, the development within the soil is often perceived as being the essence of understanding soil behavior, but in some areas of Earth, surface processes are at the heart of soil behavior, and that is certainly the case Iceland.

Sandy environments, which generally are referred to as *aeolian environments*, have a considerable share in Earth's terrestrial systems. Sandy areas are also a source of dust that can be blown over enormously long distances, to produce aeolian sediments. While the sandy environments are hostile to life and among the least populated areas on Earth, aeolian deposits provide the media for soils of some of the most fertile areas on Earth, including fertile agricultural lands in the USA, Europe, Asia, and Latin America. These deposits, often referred to as *loess*, are of various ages and composition, with a pronounced proportion of loess deposits produced at the margin of the Pleistocene Continental glaciers. Current aeolian processes and dust production continue to have an effect on Earth's ecosystems, both negative and positive. Dust has harmful effects human health, and affects such factors as snow-melt, atmospheric radiation, and climate (Field et al. 2010). However, dust can also bring in fresh parent materials that can help rejuvenate fertility of ecosystems (Vitousek et al. 2003; Pelzer et al. 2010). Iceland is an active laboratory to study the loess production of the past as well as the effect of aeolian deposition in the present.

The nature of the aeolian environments can be considered a special field of scientific study; a field that involves a range of scientific disciplines including soil science, ecology, geomorphology, meteorology, geology, and engineering. Several textbooks have been devoted to the aeolian environments and processes (e.g., *Desert Geomorphology* by Cooke et al. (1993) and *Aeolian Sand and Sand Dunes* by Pye and Tsoar (1990), and journals such as the *Journal of Arid Environments* and *Aeolian Research* are devoted to such studies. Welland (2009) provided an interesting insight into sand and sandy environments, past and present in his book titled *Sand. The Never-Ending Story*.

An overview of the various desert ecosystems was presented in Chap. 4, and on the soils of the deserts termed Vitrisols in Chaps. 6–9. The desert soils are classified as Vitricryands according to Soil Taxonomy, but a special class is designated for these soils in Iceland for many reasons (see Sect. 6.5). In this chapter, the attention is given to the active aeolian environments of Iceland.

When considering the aeolian environments, it is important to realize how wind moves soil and sediment particles. They are moved by three different modes. The collision of moving particles causes new ones to be lifted up; they gain momentum with the wind before colliding with several others when they hit the ground. This bouncing movement of materials is *saltation*, which is the most destructive form of wind erosion movement. Finer particles become airborne as dust, in *suspension*. Larger grains (limit often set at about 1 mm) are hit by the saltating materials and may be moved along the surface this way, which is termed *creep*. Continuous wind erosion leads to sorting, the dust particles are gradually lost while to proportion of saltation materials increases.

In active wind erosion areas, the majority of particles are moved by saltation. The saltation layer, where the materials are bouncing forward, usually extends to 20–30 cm height. In Iceland, the saltation layer easily reaches >100 cm height, and more coarse materials (several millimeters in diameter) are transported because of extreme wind speeds and because some of the tephra (volcanic ash) materials are of light density.

O. Arnalds, *The Soils of Iceland*,
World Soils Book Series, DOI 10.1007/978-94-017-9621-7_11
© Springer Science+Business Media Dordrecht 2015

Fig. 11.1 Sandy surface in NE Iceland. This surface is very unstable and undergoes severe wind erosion during high intensity storms

11.2 Icelandic Sand Surfaces and the Origins of the Sand

11.2.1 Extent

Iceland has about 20,000 km^2 of sandy desert surfaces. Their distribution is shown in Fig. 11.2. The desert surfaces were mapped with the survey of soil erosion in Iceland, which was presented in the book "Soil Erosion in Iceland" (Arnalds et al. 1997, 2001a). Arnalds and Kimble (2001) studied the general pedological characteristics of the deserts and the aeolian environments were reviewed by Arnalds et al. (2001b, 2012). The mapping by Arnalds et al. (2001a) employed erosion severity scores from 0 (no erosion) to 5 (very severe erosion), with scores 4 and 5 for sandy deserts representing active surfaces, but 3 representing occasionally active (see also Chap. 12). The scores were based on evidence of sediment transport in the field. The mapping of sandy surfaces differentiated between three types of geomorphic surfaces: *sand-fields* (sandur), *sandy lag-gravel* and *sandy lava* surfaces (Fig. 11.3a–c).

11.2.2 Sand-Fields

The sand-fields represent the most unstable sandy environments. These surfaces have relatively small proportion of coarse fragments and are often characterized by level surfaces with minimal surface roughness. The sand-fields vary considerably in size, from few hectares to thousands of square kilometers. Examples of large sand-fields include Dyngjusandur, north of Vatnajökull, Skeiðarársandur, south of Vatnajökull, Mýrdalssandur, southeast of Mýrdalsjökull, and Mælifellssandur, north of Mýrdalsjökull (Fig. 11.4). Geomorphology and formation of some of the sand-flats were discussed by Kjær (2004) for Mælifellssandur and Mýrdalssandur, Krüger (1997) for Mýrdalssandur, and Russell et al. (2001) for Skeiðarársandur. Some aspects of the glacial margins in association with extreme floods (jökulhlaups) were discussed by Maizels (1997), and the Icelandic aeolian conditions are a prominent part of overviews on glaciogenic dust provided by Bullard (2013).

The sediments are brought on to most of the sand-fields from underneath the glaciers with glacial rivers. There are two main modes of sand accumulation:

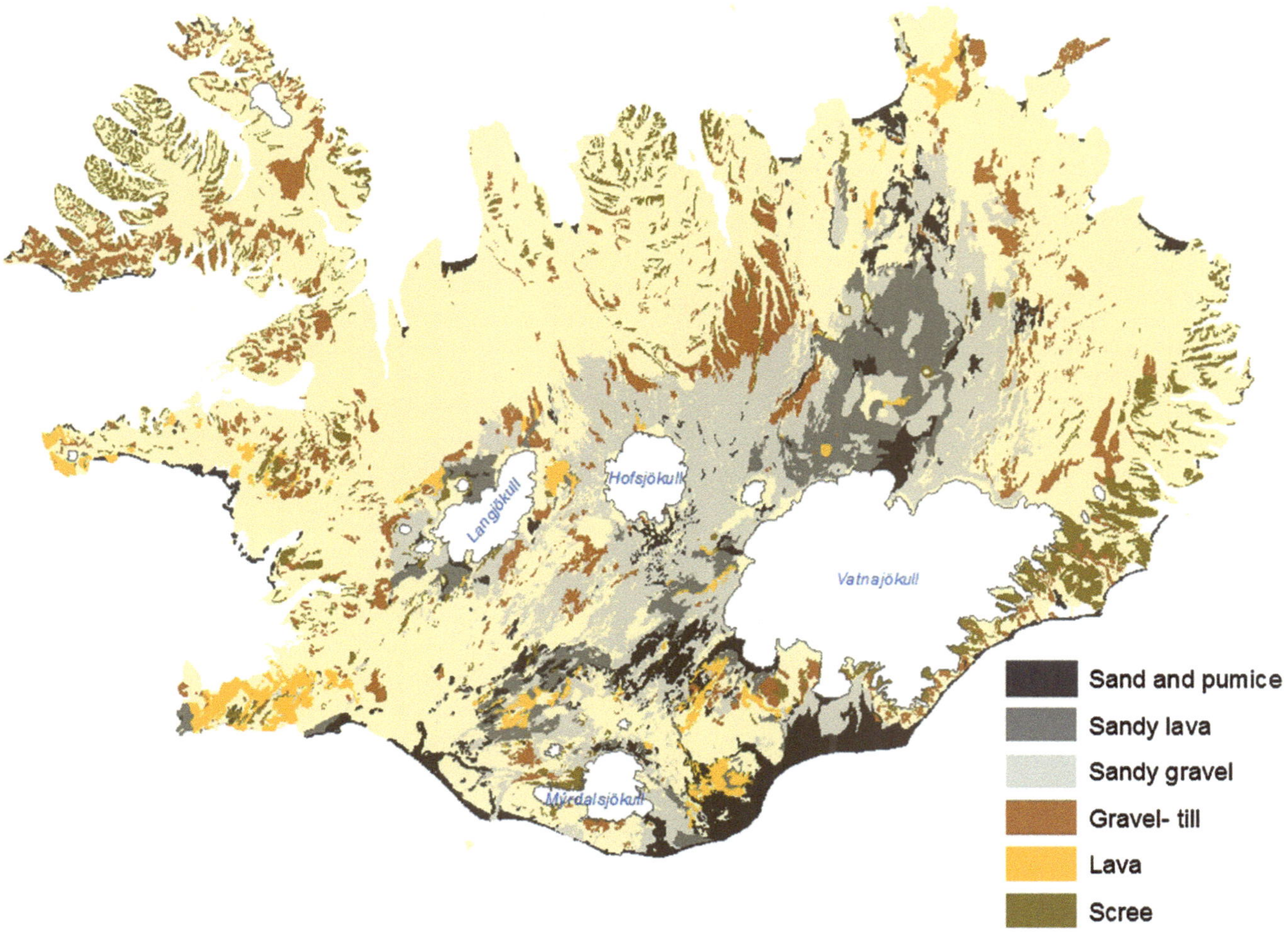

Fig. 11.2 Sandy deserts in Iceland, shown in *gray* to *black colors*. They cover large portions of the south coast and glacial margins of the active volcanic zone from Mýrdalsjökull glacier to areas northeast of Vatnajökull glacier. Map based on AUI soil erosion database, prepared by Sigmundur Helgi Brink/© SHB/OA, AUI

(i) *Temporary (often daily) floods during warm days or rains* cause the water levels of the glacial streams to rise, often with heavy sediment load. The water level recedes every night with cooler temperatures, or after the rain (Fig. 11.4). Consequently loose sediments are left on the ground, often consisting chiefly of coarse silt and fine sand (Fig. 11.5). The sediment load tends to be high in many of the glacial streams. The reason is that active volcanoes are underneath most of these major glaciers, and the rocks are often young, poorly consolidated volcanic materials, thus giving little resistance to abrasion by the glacier. In addition, the glaciers also contain tephra layers at depth that add to the fluvial sediment load. The tephra has often spread from the deposition areas over larger areas with wind.

(ii) *Large floods caused by volcanic eruptions or sudden release form subglacial water reservoirs.* Volcanic eruptions under glaciers cause spectacular explosive events when the hot magma meets the melt water of the glacier. Extreme floods can occur during such eruptions, which temporarily can equal some of the largest rivers on Earth with flow of >200,000 $m^3 s^{-1}$ (Eliasson et al. 2007). The 1996 flood in Skeiðará (south of Vatnajökull) peaked at about 50,000 $m^3 s^{-1}$ with a total sediment transport of about 180 million m^3 (Russel et al. 2006). Vatnajökull also has some sub-glacial lakes that accumulate melt water from thermal areas under the glaciers. When sufficient water has accumulated, the glacier gives in and water rushes out from underneath the glacier (see Björnsson and Palsson 2008; Björnsson 2009). Floods of this kind are in part responsible for many of the sandy areas in Iceland. Sandy deserts subsequently spread from the main deposition areas.

While most of the sediments are deposited at the margins of the glaciers, the glacial rivers carry a heavy sediment load further downstream. Where these rivers flow over relatively level land, with reduced speed of flow, sedimentation can be

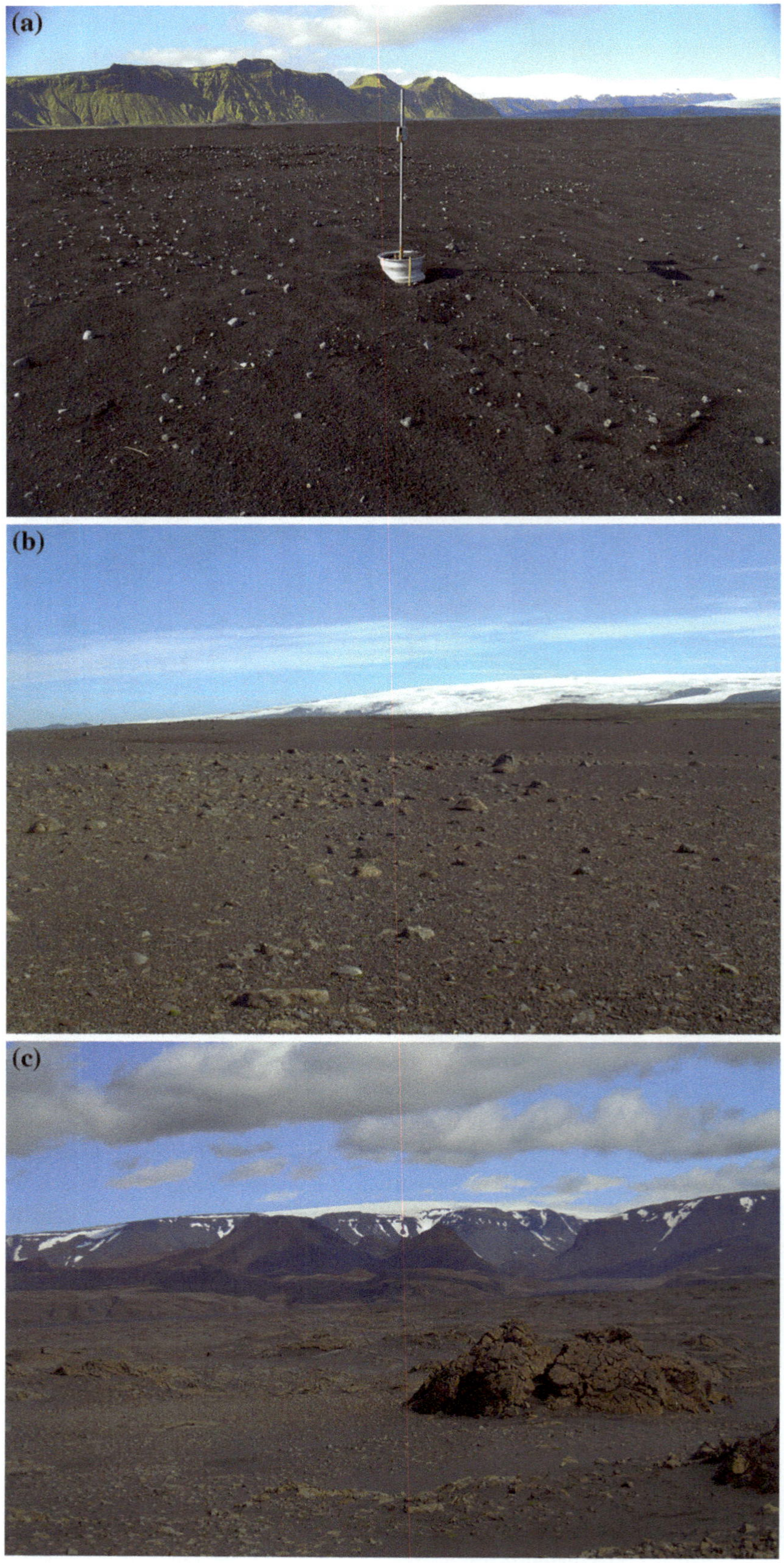

Fig. 11.3 Examples of sandy surfaces in Iceland. **a** Sand-field at Mýrdalssandur, South Iceland. The wheel-rim was previously level with the sandy surface, but wind erosion has removed enormous amounts of materials with deflation of 20–30 cm at the location. **b** Sandy-lag gravel in the central highlands. **c** Sandy lava surface south of Langjökull glacier. The sand deposits of glacio-fluvial origin have gradually filled up and abraded the lava surface. Gravel fragments from the lava also on the surface

Fig. 11.4 Mælifellssandur, a sand-field north of Mýrdalsjökull. The sand-fields are flooded during the day, charging the surface with silty materials. The water channels change frequently. Dust storms are extremely common within this area during summer (often daily), but less frequent during winter when the area is usually covered with snow. Some patches can be lowered (deflation) by several cm in single storms while some remain moist from the rivers and some areas accumulate sand by saltation

expected. These areas can gradually develop into sand-fields that may advance over the surrounding areas. This is common along the mighty Jökulsá á Fjöllum (Northeast highlands), particularly after volcanic induced floods, and along the Skaftá river in South Iceland, both draining Vatnajökull glacier.

The massive sediment load of the glacial rivers creates sandy coastlines, which often tend to be unstable aeolian environments. It has been estimated that Icelandic rivers deliver 60–70 million tons annually to the oceans (research reviewed by Gislason 2008). Gislason (2008) further noted that extreme flood events associated with volcanic eruptions and emptying of sub-glacial water reservoirs may bring the annual average up to 120–140 million tons. This has resulted in large sandy coastal areas such as the southern shoreline and Héraðssandur in East Iceland and Jökulsársandur in NE Iceland.

Some of the mapped sand-fields include areas where volcanic ash has been deposited and subsequently reworked by the aeolian and fluvial processes, such in the area around Mt. Hekla in South Iceland (Thorarinsdottir and Arnalds 2012).

The soils of the sand-field areas classify as Vitrisols, but Vitric Andosols or Vitricryands according to the WRB and Soil Taxonomy (respectively), even though these deposits are often very fresh (see Chap. 6 on Classification).

11.2.3 Sandy Lag-Gravel

The sandy lag-gravel surfaces are formed on glacial till left by the Pleistocene glacier, raised shorelines, and alluvial surfaces. These surfaces often have accumulated appreciable amounts of aeolian sand (often 0.1–1 mm year^{-1}) and volcanic tephra, trapped by the coarse fragments on the surface. Frost heave elevates the coarse fragments every winter, maintaining the gravelly surface. This is causing the surface to gradually rise, accumulating, sandy layers below the surface. Sandy lag-gravel surfaces become unstable during dry high intensity winds, causing massive wind erosion and dust production. Much lower wind intensities are required for dust production on the unstable sand-fields described above. Losses of materials from sandy lag-gravel surfaces

Fig. 11.5 Loose silty glacio-fluvial sediments on Dyngjusandur, recently deposited during temporary flooding caused by warm summer weather. Wind erosion at beginning stages early in the day, but some moisture still remains in the deposits close to the photographer. This area is perhaps the most active aeolian dust source in Iceland. *Photo* © Sveinn Runólfsson

Fig. 11.6 Wind abraded rock, termed ventifact (*center of picture*), in an Icelandic desert. *Photo* © Fanney Osk Gisladottir

during wind erosion events counteract the aeolian sediment accumulation. Ventifacts, which are wind abraded rocks, are common on sandy lag-gravel surfaces (Fig. 11.6). Sandy lag gravel is the most common sandy surface in Iceland, with a total of 12,910 km^2, but about 7,500 km^2 of these have a very unstable surfaces (erosion score 4 and 5; Table 11.1). Many of the sandy lag-gravel areas were previously covered with Andosols, with full vegetation cover, but have become desertifed after the Settlement.

Table 11.1 Active aeolian surfaces in Iceland

Surface type	Erosion Score			Total
	3	4	5	
	km^2			
Sandur (sand-fields)	318	1,087	2,828	4,233
Sandy lag-gravel	5,407	6,217	1,286	12,910
Sandy lava	1,366	1,757	1,620	4,743
Total	7,091	9,061	5,734	21,886
Total 4 + 5		14,795		

Erosion scores represent erosion activity with *3* considerable; *4* severe; *5* very severe. Based on "Soil Erosion in Iceland" (Arnalds et al. 2001a). The combined areas of erosion score 4 and 5 represent the active aeolian areas (14,795 km^2)

11.2.4 Sandy Lava Surfaces

There are large areas of lava surfaces dating from the Holocene, formed after the last major glaciation (11,700 km^2; Johannesson and Saemundsson 2009). Some of these lavas have become fully vegetated with fertile Brown Andosols underneath. A large part of the lavas are barren, with accumulation of silty and sandy deposits in depressions from both aeolian processes and direct deposition of tephra during volcanic eruptions. Lava surfaces with silty/sandy deposits cover more than 4,700 km^2 (Table 11.1). The combination of rock outcrops and sandy deposits creates an interesting combination of Leptosols and Sandy Vitrisols. The surface roughness of the lavas is an important factor determining the fate of the sand. Relatively flat "pahoehoe" lavas are more common than the rougher "aa" lavas, but there are also intermediates ("rubbly pahoehoe") lavas (see Thordarson and Höskuldsson 2008). The rougher lavas can accumulate immense amount of sand before they become unstable sand-fields, often providing a temporary partial shelter from sand for the downwind areas. Extensive sandy lava areas are found within the Mt. Hekla lava-fields, which have received large amount of tephra, but there are also large lava-fields in Northeast Iceland (Ódáðahraun) filled with sand, and a large proportion of the sand has been blown long distances over the lavas from the sources closer to Vatnajökull glacier or the Jökulsá á Fjöllum glacial river. Unstable sandy lava areas (erosion score 4 and 5) are about 3,400 km^2.

11.3 Composition

Considering, the geology of Iceland, it is not surprising that the majority of the sand is comprised of basaltic volcanic glass, often originating from hyaloclastic ridges formed during volcanic eruptions under the glaciers (see Baratoux et al. 2011) or tephra deposited over the landscapes during volcanic eruptions. Typical low Si and high Fe basaltic sands are at the margins of Vatnajökull glacier (including the Dyngjusandur and Skeiðarársandur dust-plume sources) and the Mýrdalsjökull glacier (Mýrdalssandur and Mælifellssandur dust-plume sources; see Baratoux et al. 2011; Oladottir et al. 2008, 2011; Dagsson-Waldhauserova et al. 2014). Andesitic glass is also common, especially near Mt. Hekla (numerous andesitic eruptions over the past 1,000 years). Few rhyolitic tephra layers from large eruptions have wide-spread distribution in soils, such as the large rhyolitic Hekla tephra layers (see Chap. 3 on Geology). However, rhyolitic deposits have limited surface extent except close to few volcanoes.

The active sand surfaces within the Hagavatn area south of Langjökull (Figs. 11.7 and 11.8), differ from other sandy deposits in Iceland, because this sand is largely crystallized basalt (Baratoux et al. 2011; Mangold et al. 2011). It seems that a glacier has advanced over few thousand years old lava, leaving abraded lava particles behind as it retreats, and adding to the pile created by the glacial rivers coming from underneath the glacier (see Gisladottir et al. 2005; Mangold et al. 2011).

Volcaniclastic aeolian environments are reported in many areas on Earth, but are not nearly as common as quartz-rich, carbonatic, or clay/silt aggregate aeolian materials (Edgett and Lancaster 1993). Iceland has, by far, the largest such volcaniclastic fields (see Arnalds et al. 2001b). These sand-fields have been used as analogs for the desert landscapes and processes on the planet Mars (e.g., Baratoux et al. 2011; Mangold et al. 2011). The research by Mangold et al. (2011) showed that wind erosion in Iceland can be effective in sorting of minerals because of difference in their shape, grain size, and density.

11.4 The Redistribution of the Materials

Wind erosion, the redistribution by wind of loose materials, is the subject of research all over the world, on agricultural fields, rangelands, and within the sandy deserts. However, there are several ways to report quantification of wind erosion. The most common response value, which is employed by the widely cited Wind Erosion Equation and similar equations (see e.g., Skidmore 1994) is tons of soils lost from each hectare over given time (tons ha^{-1} year^{-1}). This value often has limited meaning in sandy environments where both sedimentation and transport can occur simultaneously. Another method is to measure transport over a 1 m wide line, often during a given time (sec, min, hrs, days, years). We in Iceland have commonly employed the response value kg m^{-1} h^{-1} (Arnalds et al. 2012), or kg of materials transported over 1 m wide line in 1 h (see Fig. 11.9 for equipment and methods). Similar response value used in the literature is

Fig. 11.7 The Hagavatn dust plume source in front of the Hagfellsjökull glacier. The shore of the glacial lake with unstable water level, and the fluvial areas around the glacial river feeding the lake are a major source of dust production. The river flows over a drained lake bed with fine glacial deposits. More coarse materials are saltated long distances to the south (*right*) over the lava-fields, creating sandy lava surface. During severe storms, the sand-flats south of the river and the lake also become dust sources. Mountain peaks at the *center* of the photo are the Jarlhettur formation, formed by a fissure eruption under glacier during the last glaciation

Fig. 11.8 A concentrated dust-plume rising from the Hagavatn dust source in July 2009. The photograph is taken from a 40 km distance from the source. *Photo* © Asa L. Aradottir, AUI

kg m^{-2} or even g cm^{-2} (e.g., van Donk and Skidmore 2001; Skidmore et al. 1994). It is also common to measure deflation rates (e.g., mm $year^{-1}$) or the lowering of the surface (e.g., Seppälä 2004), which has the same limitation as tons ha^{-1} when removal and deposition occur within the same area. Dust production is often measured in tons per unit area (e.g., tons km^{-2}) which also can be time or event referenced (tons km^{-2} $year^{-1}$ or per storm). These different

Fig. 11.9 Wind erosion measurement equipment in the field. On the *left* are 5 Fryrear samplers (BSNE samplers; Fryrear 1986), from 10 to 120 cm height. In the *center* are SENSIT wind erosion sensors at three different heights, each giving pulse when a soil grain hits a small piezo-electric plate. The pulses are gathered in a data-logger, together with various climatic parameters, and the data can be accessed by a telephone (remote research site)

methods sometimes cause confusion in the literature, which is why this is brought up here.

When silty and sandy materials have been deposited by a river or during eruptions, wind erosion begins a sorting process, with the finer materials lost as dust and the larger materials left behind, but with ever increasing proportion of saltation materials downwind from the source. This sorting effect is well expressed in the vicinity of areas of fluvial deposition at the glacial margins within plume areas (see later in the chapter) such as the Hagavatn area and at the Mýrdalssandur sand-field. Materials can be blown long distances (tens of kilometers) from the source areas, forming a kind of rivers of sand ("sand-paths"), causing severe wind erosion far from the point of origin (Fig. 11.10). Sand seems to be blown >100 km in the Ódáðahraun area in Northeast Iceland (Arnalds 1992).

11.5 Wind Erosion Rates in Iceland

What are the rates of surface transport of aeolian materials in Iceland? It seems that common values for sandy areas range between 500 and 3,000 kg m^{-1} $year^{-1}$; thus, ½–3 tons are blown over a 1 m wide transect each year. However, transport caused by storms within the most unstable areas is even greater: >500 kg m^{-1} is common during single storms (Thorarinsdottir and Arnalds 2012; Arnalds et al. 2012, 2013). This enormous transport has very severe consequences when it reaches fully vegetated systems, but such destruction is a major factor in the collapse of Icelandic ecosystems ("advancing sand-fronts", see Chap. 12). Traveling over some of the sand-fields during sandstorms can be quite detrimental, as visibility is near zero, paint on automobiles can be scraped entirely off on the side facing upwind, the windows become opaque and even broken by the large particles moved by the wind.

Volcanic eruptions can leave vast amounts of loose materials on the surface. Fully vegetated systems can adsorb considerable volcanic ash deposition without severe damage or subsequent redistribution, but if the ash falls on relatively barren land, it becomes unstable and is subjected to erosion both by wind and water. Wind erosion measured after the 2010 Eyjafjallajökull eruption rates among the most violent sandstorms ever reported in the scientific literature, with >11 tons of materials blown over a 1 m wide transect in a single storm (Arnalds et al. 2013) with severe dust problems in extensive areas after the eruption (Thorsteinsson et al. 2012).

11.6 The Dust Hotspots: Sandy Areas and Dust Plume Areas

Dust coming from wind erosion areas is the major source for the parent materials of Icelandic soils. Dust can be generated from most of the sandy surfaces during the high intensity storms, but the amount of finer particles varies in the sediments, hence, the sorting effect caused by downwind

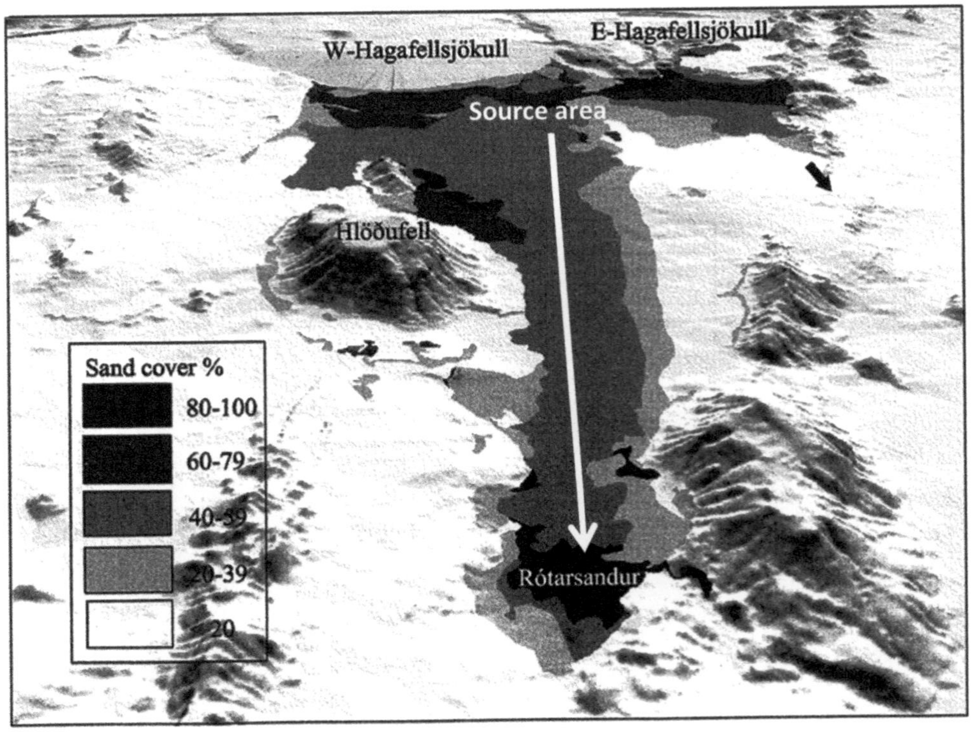

Fig. 11.10 River of sand south of the Hagavatn area. The source area are frequently flooded by glacial streams loaded with sediments, yet on flat land, hence lack of channeling. Sand is blown southwards >16 km and deposited at Rótarsandur sand-field. The river was slowly being formed as the lava terrain gradually filled up by sand. The source area is one of the major dust plume sources in Iceland. Adapted from Gisladottir et al. (2005)

movement of the sand from its source. This has an influence on the dust production of the sandy areas. The areas that generate dust most frequently and in the largest quantities have been termed *plume areas* to distinguish them from other sandy areas (Arnalds 2010). A map showing the major plume areas are presented in Fig. 11.11.

Weather reports from weather stations in Iceland show that >130 dust events occur each year, and many of the (5–15) can be considered major with dust generation exceeding 500,000 tons in each storm (based on Arnalds et al. 2014; see also Dagsson Waldhauserova et al. 2013). An example of a major storm from Dyngjusandur dust-plume source in North Iceland (see Fig. 11.11) is presented in Fig. 11.12a. Arnalds et al. (2014) estimated that 300–800 tons of aeolian materials are blown during each storm of this size. The dust plume extends >200 km from the source on the image. Figure 11.12b illustrates dust being blown from South Iceland, but the sources include recently deposited volcanic ash.

11.7 Quantification of Aeolian Sedimentation in Iceland and Implications for Soils and Ecosystems

Arnalds (2010) presented a map of dust sedimentation rates for dust deposition in Iceland (Fig. 11.13). The map is an approximation based on published research and the Agricultural University Soil Database for soil thickening rates of Brown Andosols. The results show that sedimentation rates range from <50 to >250 g m^{-2} according to the map, but the highest values approach about 1,000 g m^{-2}, which is among the highest reported in the world literature (Lawrence and Neff 2009). A modified version of this map (more deposition classes) was published by Gunnarsson et al. (2014) and it was extended to oceanic areas by Arnalds et al. (2014). There is a huge amount of dust that is created by the Icelandic dust sources each year, which has been estimated of the order of 30–40 million tons each year (Arnalds et al. 2014). Most of this dust is redeposited on land but a large fraction (5.5–13.8 million tons) is deposited on sea, which most likely affects the primary production in the oceans around Iceland (Arnalds et al. 2014).

The steady aeolian accumulation results in the gradual rise of the surface, which has occurred since vegetation began to trap the sediments after the retreat of the Pleistocene glacier. Since then, thicknesses from few decimeters to >2 m of sediments have been accumulated, providing parent materials for Icelandic soils. This is, of course, an analog for loess deposition elsewhere in the world, but volcanic loess materials are, however, rather rare. It should be emphasized that most soil profiles show rapid increase in sedimentation rates after the Settlement (874 AD), as is discussed in Chap. 12. A part of that increase results from redistribution of existing soils (mostly Brown Andosols), which have high component of silty materials with suspended materials (dust)

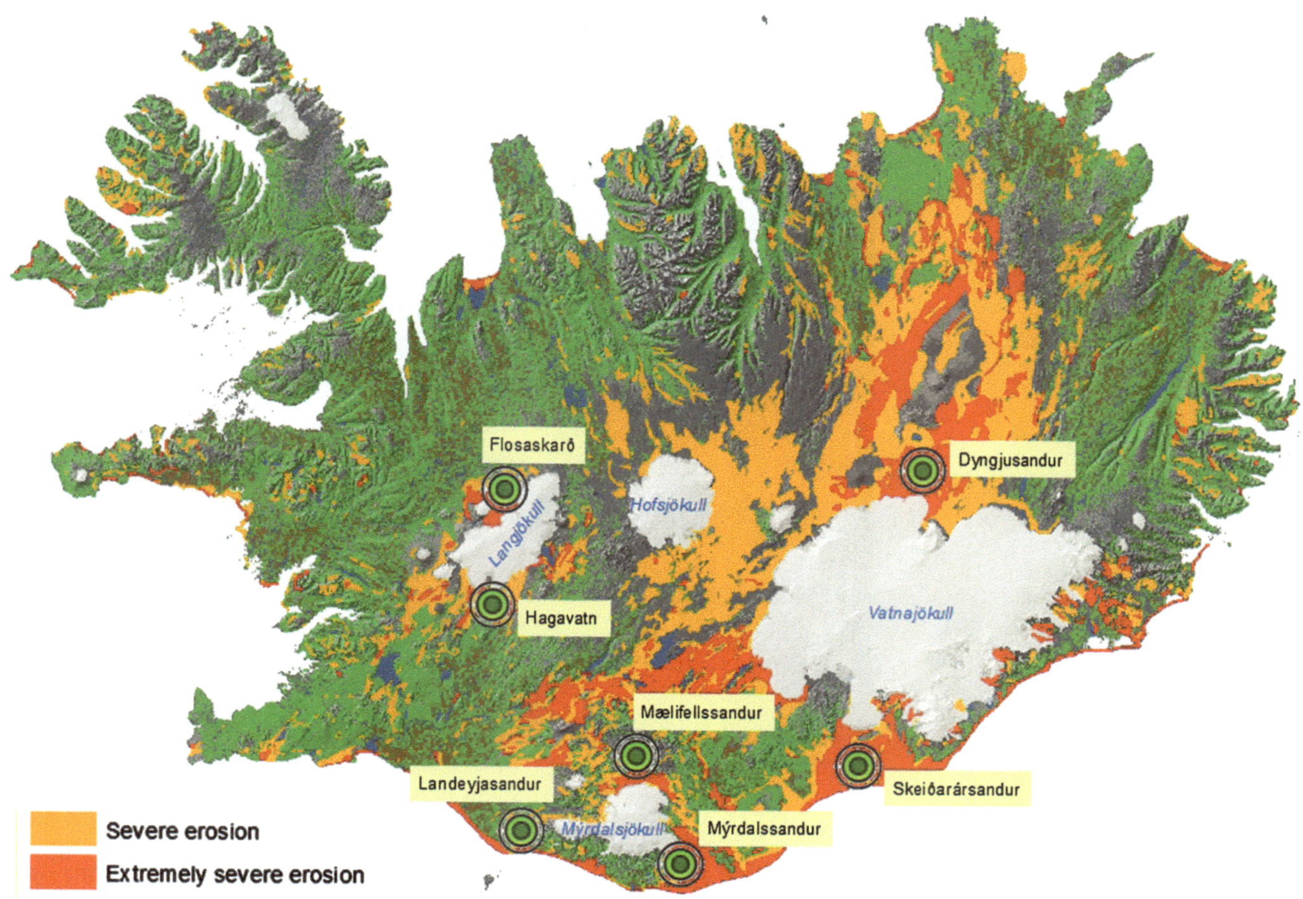

Fig. 11.11 Main plume areas in Iceland (Arnalds 2010). These plume areas, major sources of aeolian materials in Iceland, are located at the margins of the glaciers and along glacial rivers. There are smaller additional plume sources. Red colors indicate sandy deserts and erosion severity (score), based on Arnalds et al. 2001a, but the sandy deserts are also unstable and become active dust sources during the most intense storms, but much less frequently than the plume sources indicated by the *dots* in the figure

being a major component the aeolian materials when such soils are subjected to wind erosion (see Chap. 12).

The author of this book has emphasized the influence of the dust and tephra deposition on Icelandic ecosystems, which is part of the underlying principles for the classification of the soils. High deposition rates result in lower organic carbon and clay contents (per depth increment) but high pH values, while low deposition rates enhance carbon levels and low pH. Clay contents are highest at intermediate stages (see Chaps. 6–9). Birds are a sensitive measure of ecosystem condition, generally being at the top of the food chain. It is therefore interesting to note that Gunnarsson et al. (2014) showed that the ecosystem fertility, as indicated by bird numbers in dryland and wetland habitats, was closely linked to the amount of aeolian deposition; more deposition enhances ecosystem fertility.

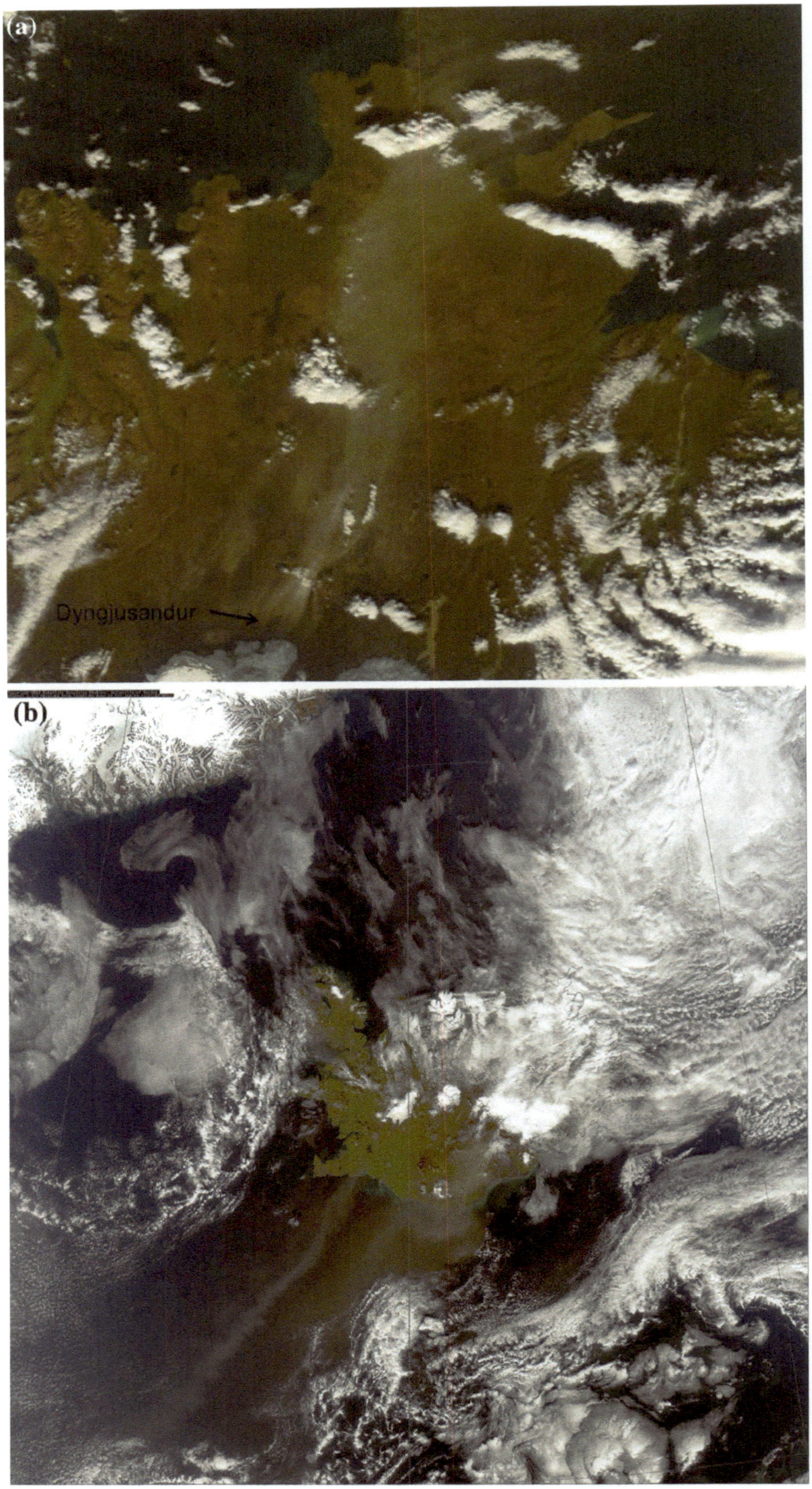

Fig. 11.12 **a** A dust plume extending from the Dyngjusandur dust source over NE Iceland and into the oceans north of the island. (see Dagsson-Waldhauserova 2013; Arnalds et al. 2014). **b** Dust being blown from the major dust sources in South Iceland into the North-Atlantic Ocean. The dust also includes recent unstable volcanic ash from the 2011 Grímsvötn eruption. *Both images* ©NASA (MODIS)

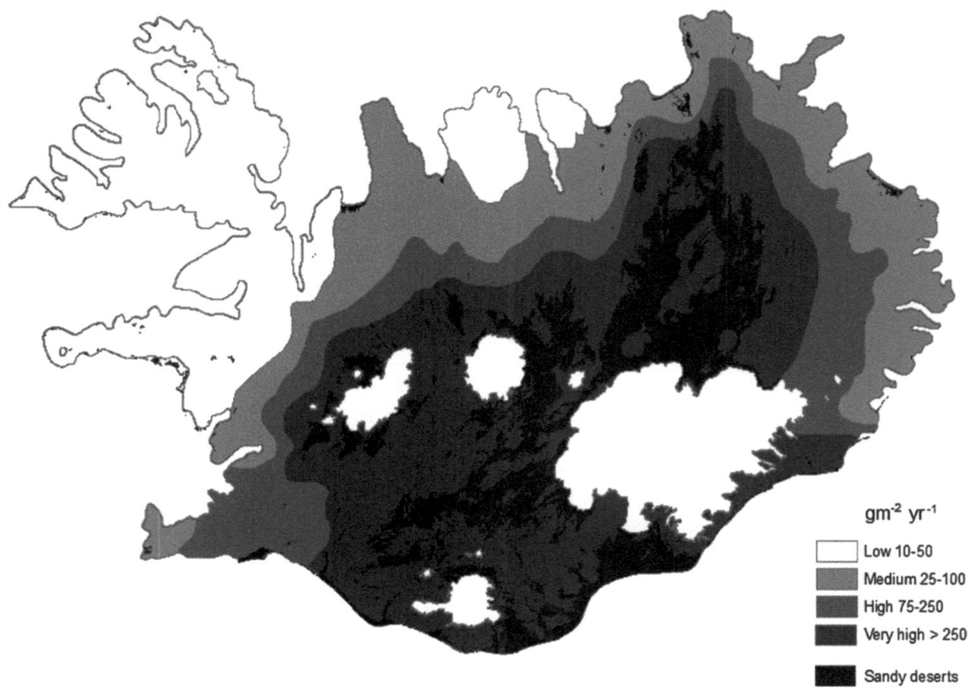

Fig. 11.13 Aeolian deposition rates in Iceland. The *black areas* under the *shading* represent the sandy deserts in Iceland. The deposition rates also include tephra deposition from volcanoes. Map published by Arnalds (2010)

References

Regarding punctuation and Icelandic characters in citations: See note on punctuation in the Preface

Arnalds O (1992) Pathways of aeolian sand-drift in Northeast Iceland (Sandleiðir á Norðausturlandi). Græðum Ísland (Icelandic SCS yearbook) IV:145–149. Soil Conservation Service of Iceland (in Iceland)

Arnalds O (2010) Dust sources and deposition of aeolian materials in Iceland. Icel Agric Sci 23:3–21

Arnalds O, Kimble J (2001) Andisols of deserts in Iceland. Soil Sci Soc Am J 65:1778–1786

Arnalds O, Thorarinsdottir EF, Metusalemsson S, Jonsson A, Gretarsson E, Arnason A (1997) Soil erosion in Iceland (Jarðvegsrof á Íslandi). Landgraaedsla ríkisins og Rannsóknastofnun landbúnaðarins, Reykjavík (in Icelandic)

Arnalds O, Thorarinsdottir EF, Metusalemsson S, Jonsson A, Gretarsson E, Arnason A (2001a) Soil erosion in Iceland. Soil Conservation Service and Agricultural Research Institute, Reykjavik (original edition in Icelandic 1997)

Arnalds O, Gisladottir FO, Sigurjonsson H (2001b) Sandy deserts of Iceland: an overview. J Arid Environ 47:359–371

Arnalds O, Gisladottir FO, Orradottir B (2012) Determination of aeolian transport rates of volcanic soils in Iceland. Geomorphology 167–168:4–12

Arnalds O, Thorarinsdottir EF, Thorsson J, Dagsson-Waldhauserova P, Agustsdottir AM (2013) An extreme wind erosion event of the fresh Eyjafjallajökull 2010 volcanic ash. Sci Rep 3:1257. doi:10.1038/srep01257

Arnalds O, Olafsson H, Dagsson-Waldhauserova P (2014) Quantification of iron-rich volcanogenic dust emissions and deposition over ocean from Icelandic dust sources. Biogeosci Discuss 11:5941–5967

Baratoux D, Mangold N, Arnalds O, Bardintzeff J-M, Platevoet B, Grégorie M, Pinet P (2011) Volcanic sands of Iceland—diverse origins of aeolian sand deposits revealed at Dyngjusandur and Lambahraun. Earth Surf Process Landf 36:1789–1808

Björnsson H (2009) Jöklar á Íslandi (Icelandic Glaciers). Opna, Reykjavik (in Icelandic)

Björnsson H, Palsson F (2008) Icelandic glaciers. Jökull—Icel J Earth Sci 58:365–386

Bullard JE (2013) Contemporary glacigenic inputs to the dust cycle. Earth Surf Process Land 38:71–89

Cooke R, Warren R, Goudie A (1993) Desert geomorphology. UCL Press, London

Dagsson-Waldhauserova P, Arnalds O, Olafsson H (2013) Long-term frequency and characteristics of dust storm events in Northeastern Iceland (1949–2011). Atmos Environ 77:117–127

Dagsson-Waldhauserova P, Arnalds O, Olafsson H, Skrabalova L, Sigurdardottir GM, Branis M, Hladil J, Skala R, Navratil T, Chadimova L, von Lowis of Menar, S Thorsteinsson Th, Carlsen HK, Jonsdottir I (2014) Physical properties of suspended dust during moist and low-wind conditions in Iceland. Icel Agric Sci 27: 25–39

Edgett KS, Lancaster N (1993) Volcaniclastic aeolian dunes: terrestrial examples and application to martian sands. J Arid Environ 25:271–297

Eliasson J, Kjaran SP, Holm SL, Gudmundsson MT, Larsen G (2007) Large hazardous floods as translatory waves. Environ Model Softw 22:1392–1399

Field JP, Belnap J, Breshears DD, Neff JC, Okin GS, Whicker JJ, Painter TH, Ravi S, Rehis MC, Reynolds RL (2010) The ecology of dust. Front Ecol Environ 8:423–430

Fryrear DW (1986) A field dust sampler. J Soil Water Conserv 41:117–121

Gisladottir FO, Arnalds O, Gisladottir G (2005) The effect of landscape and retreating glaciers on wind erosion in South Iceland. Land Degrad Dev 16:177–187

Gislason SR (2008) Weathering in Iceland. Jökull—Icel J Earth Sci 58:387–408

Gunnarsson TG, Arnalds O, Appleton G, Gill JA (2014) Ecosystem recharge by volcanic dust drives large-scale variation in bird abundance (in prep)

Johannesson H, Saemundsson K (2009) Geological map of Iceland 1:600,000. Icelandic Institute of Natural History, Reykjavik

Kjær KH (2004) Architecture and sedimentation of outwash fans in front of the Mýrdalsjökull ice cap, Iceland. Sed Geol 172:139–163

Krüger J (1997) Development of minor outwash fans at Kötlujökull, Iceland. Quat Sci Rev 16:649–659

Lawrence CR, Neff AJ (2009) The contemporary physical and chemical flux of aeolian dust. A synthesis of direct measurements of dust deposition. Chem Geol 267:46–63

Maizels J (1997) Jökulhlaup deposits in proglacial areas. Quat Sci Rev 16:793–819

Mangold N, Baratoux D, Arnalds O, Bardintzeff J-M, Platevoet B, Grégoire M, Pinet P (2011) Segregation of olivine grains in volcanic sands in Iceland and implications for Mars. Earth Planet Sci Lett 310:233–243

Oladottir BA, Sigmarsson O, Larsen G, Thordarson T (2008) Katla volcano, Iceland: magma composition, dynamics and eruption frequency as recorded by Holocene tephra layers. Bull Volcanol 70:475–493

Oladottir BA, Sigmarsson O, Larsen G, Devidal J-L (2011) Provenance of basaltic tephras from Vatnajökull subglacial volcanoes, Iceland as determined by major- and trace-element analyses. Holocene 21:1037–1048

Pelzer DA, Wardle DA, Allison VJ, Baisden WT, Bardgett RD, Chadwick OA, Condron LM, Parfitt RL, Porder S, Richardson SJ, Turner BL, Vitousek PM, Walker J, Walker LR (2010) Understanding ecosystem retrogression. Ecol Monogr 80:509–529

Pye K, Tsoar H (1990) Aeolian sand and sand dunes. Unwin Hyman, London

Russel AJ, Knight PG, Van Dijk TAGP (2001) Glacier surging as a control on the development of proglacial, fluvial landforms and deposits, Skeiðarársandur, Iceland. Global Planet Change 28:163–174

Russell AJ, Roberts MJ, Fay H, Marren PM, Cassidy MJ, Tweed FS, Harris T (2006) Icelandic jökulhlaup impacts: Implications for ice-sheet hydrology, sediment transfer and geomorphology. Geomorphology 75:33–64

Seppälä M (2004) Wind as a geomorphic agent in cold climates. Studies in polar research. Cambridge University Press, Cambridge

Skidmore EL (1994) Wind erosion. In: Lal R (ed) Soil erosion research methods. Soil and Water Conservation Society, Ankeny, pp 265–293

Skidmore EL, Hagen LJ, Armbrust DV, Durar AA, Fryrear DW, Potter KN, Wagner LE, Zobeck TM (1994) Methods for investigating basic processes and conditions affecting wind erosion. In: Lal R (ed) Soil erosion research methods. Soil and Water Conservation Society, Ankeny, pp 295–330

Thorarinsdottir EF, Arnalds O (2012) Wind erosion of volcanic materials in the Hekla area, South Iceland. Aeolian Res 4:39–50

Thordarson T, Höskuldsson A (2008) Postglacial volcanism in Iceland. Jökull—Icel J Earth Sci 58:197–228

Thorsteinsson T, Jóhannsson T, Stohl A, Kristiansen NI (2012) High levels of particulate matter in Iceland due to direct ash emissions by the Eyjafjallajökull eruption and resuspension of deposited ash. J Geophys Res 117: B00C05. doi:10.1029/2011JB008756

van Donk SJ, Skidmore EL (2001) Field experiments for evaluating wind erosion models. Ann Arid Zone 40:283–302

Vitousek P, Chadwick O, Matson P, Allison S, Derry L, Kettley L, Luers A, Mecking E, Monastra V, Porder S (2003) Erosion and rejuvenation of weathering-derived nutrient supply in an old tropical landscape. Ecosystems 6:762–772

Welland M (2009) Sand. The never-ending story. University of California Press, Berkeley

12 Collapse, Erosion, Condition, and Restoration

12.1 Collapse

Man-induced ecosystem degradation in Iceland has had more dramatic consequences than in most other countries of the northern hemisphere (Fig. 12.1). Poorly vegetated land now covers >40 % of the country, with the remaining ecosystems in a much poorer state compared to optimal conditions (e.g., Thorsteinsson et al. 1971; Arnalds 1987, 1988; Arnalds and Aradottir 2011). Woodlands were reduced from about 25 % to almost total elimination about 100 years ago (Sigurdsson 1977; Hallsdottir 1995; Aradottir and Arnalds 2001; Aradottir and Eysteinsson 2005). A large proportion of the Icelandic wetlands were drained during the last century (Gardarsson et al. 2006). The Icelandic ecosystem collapse is often cited with other examples of disastrous human impacts, such as in Jared Diamond's book *Collapse* (2005), Montgomery's book *Dirt* (2007), and the Icelandic example is given considerable attention in *Desertification, Land Degradation, and Sustainability* by Imeson (2012). Recent Icelandic natural history is not only about destruction and collapse; it is also about survival of man. Streeter et al. (2012) stated that "Iceland occupies a position of precarious survival, defined by not becoming extinct, like Norse Greenland, but having endured, sometimes by the narrowest of margins." A comprehensive overview of the nearly 1,140 years of human history of Iceland, emphasizing its location as a marginal environmental area, was published by Karlsson (2000).

Degradation of ecosystems on Earth is a severe threat to soil resources and is therefore one of the most important aspects of soil science and ecology. Large-scale land degradation is addressed in other World Soil Book Series books published by Springer, such as in *The Soils of Chile* (Casanova et al. 2013) and *The Soils of Mexico* (Krasilnikov et al. 2013), but in addition to similarities reflected by severe degradation, these countries have a large share of volcanic soils as are found in Iceland.

There is a considerable wealth of research to give insight into the collapse of Icelandic ecosystems, which will be discussed below with references to aid those with interest in the subject. The current state of the resources, including vegetation distribution, major vegetation classes, and erosion activity is relatively well documented. Research methods and the state of the land will be described in this chapter. Finally, restoration activities in Iceland will be discussed briefly.

12.2 Resilience and Impacts: A Little History of Soils and Vegetation

12.2.1 Evidence of Ecosystem Changes

Iceland was almost entirely covered by glacier during the Pleistocene glaciation. After the retreat of the glacier, some 9,000–10,000 years ago, Iceland gradually became nearly fully vegetated, with plants species surviving the glaciation or brought to the island by wind, birds, or other means. The climate has fluctuated considerably during the Holocene, causing large-scale vegetation changes, especially in abundance of wetland species and birch woodlands. Einarsson (1968) splits the Holocene vegetation history in Iceland into periods or zones based on pollen analysis, with alternating birch and mire/wetland periods, reflecting climatic fluctuations during the period. The current epoch is the second mire period, but birch was dominant with apparent dryer conditions during two prehistoric periods (see also Hallsdottir 1995; Hallsdottir and Caseldine 2005). But the most dramatic change during the Holocene in vegetation composition and soil cover occurred with the arrival of man to Iceland.

Considerable knowledge has been gathered from multiple sources about the past riches and environmental changes in Iceland, but these sources were summarized by Arnalds (1987). These include: (i) medieval written records such as the Sagas and Annals; (ii) vegetation and soil remnants in severely degraded areas; (iii) pollen analyses; (iv) old charcoal pits and remnants in present day deserts (Fig. 12.2); (v) aeolian deposition rates; (vi) old place names; (vii) studies of lake sediments; (viii) archeological research;

O. Arnalds, *The Soils of Iceland*,
World Soils Book Series, DOI 10.1007/978-94-017-9621-7_12
© Springer Science+Business Media Dordrecht 2015

Fig. 12.1 Visitors in an Icelandic desert. Small soil and vegetation remnants to the *right* of the men. This area was previously fully covered with >1 m thick Andosol and birch forest, which was used as a source for fuel wood. The surface is now 1–2 m lower than it was prior to the destruction and vegetation is scarce. *Photo* © G.K. Johannesson

Fig. 12.2 Charcoal pits in present day deserts. A stone ring marks the place where the charcoals were found. Vegetation and soil remnants in the background. In many areas, remnants of charcoal making from cutting down birch trees are found where there is presently a desert. Most such areas can potentially be restored to former grandeur. *Photo* Andres Arnalds, ISCS

(ix) vegetation of naturally protected areas such as on islands in large rivers. There is a famous passage in *The Book of Icelanders* written by Ari Fróði (Ari the Wise) few centuries after the Settlement, which states that Iceland was covered with woodlands from the shore to the mountains at the time of the settlement. Although the validity of this statement can be questioned, Icelanders have always been aware of vast environmental changes that have occurred since the island was settled.

Many noted the poor state of the land early on, including the geographer Th. Thoroddsen and Saemundur Eyjolfsson during the nineteenth century (see Sigurjónsson 1958; Arnalds 1988). It became increasingly evident that heavy grazing the year around by sheep and wood harvesting for fuel caused widespread damage, which prompted the writing of many papers, including a landmark paper by Bjarnason (1942) on the results of overgrazing and wood cutting.

The Sagas and other written medieval documents provide many clues for the state of the land. Old place names indicate the presence of farms and forested areas where now there is severely degraded land. The meaning of some terms have changed, such as the common term "holt" which previously meant forested hill but now an unsheltered or barren area.

Erosion history has been studied in several areas in Iceland. Gudbergsson (1975) studied erosion history of the Skagafjörður district, North Iceland, mostly based on aeolian sedimentation rates, historical records, and thin sections of soils. He also published an extended review of the vegetation and land use history of NW Iceland (Gudbergsson 1996). Sveinbjarnardottir et al. (1982) and Gerrard (1985) studied erosion and landscape stability in southern Iceland. Gisladottir (1998), Gisladottir et al. (2010) have made detailed studies of the erosion history of the Krísuvík area (Coastal Southwest) and in West Iceland (Gisladottir et al. 2011). Sigbjarnarson (1969), Gisladottir et al. (2005) and more recently Greipsson (2012) studied the degradation and soil erosion south of Langjökull Glacier, where the combination of grazing, wood cutting, climate change (cold spells), and changes at the glacial margins acted to cause severe ecosystem destruction. Glacial margins, glacial rivers, and lakes with unstable water levels act as sand sources for the advancement of advancing sand fronts, with sustained sand drift enhanced by dry katabatic winds from the Langjökull glacier. All these studies provide valuable insight into the history of degradation in Iceland.

12.2.2 Aeolian Deposition Rates

Aeolian deposition rates are commonly used in Iceland to infer environmental changes, with rapid rates signaling erosion periods, but slow rates indicate periods of less aeolian activity, hence: little or less geomorphic activity associated with land degradation. The rates can be determined by measuring depths between tephra layers of known age. Sigurdur Thorarinsson pioneered the scientific field of tephrochronology and published a comprehensive review in 1961 showing that the aeolian sedimentation rates increased dramatically after the Settlement, as a result of wind erosion of soils. His aeolian deposition rate studies were followed by Sigbjarnarson (1969), who studied land degradation south of the Langjökull Glacier, Gudbergsson (1975) in North Iceland and the method was employed in a number of studies by Andrew Dugmore and coworkers (e.g., Dugmore et al. 2000). Micromorphological studies show that a large proportion of the aeolian deposits dated after the Settlement consist of redistributed soils. This is exemplified by the occurrence of redistributed rhyolitic tephra grains easily identified under the microscope and often under a hand lens (e.g., Gudbergsson 1975; Arnalds 1990; Stoops et al. 2008). The redistributed soil materials are mixed with other aeolian materials that originate from the major dust source areas discussed in Chap. 11. It was shown in Chap. 8 how organic carbon levels dropped in soils with high rate of aeolian deposition during the major advent of soil erosion during the Middle Ages.

Aeolian deposition rates remain an important part of the methodology to study past environments (e.g., Gisladottir et al. 2010; Streeter and Dugmore 2012). Care, however, must be exercised in interpreting aeolian deposition data. It is appropriate in this text about soils and the Icelandic soil environment to raise some points of caution in relation to aeolian deposition rates.

(i) Recent studies show that the most active present day sources of aeolian materials are "confined" areas of high dust productivity (Arnalds 2010; Dagsson-Waldhauserova et al. 2013; Arnalds et al. 2014). In some areas, much of the most erodible Andosols were depleted during early periods of erosion activity during the Middle Ages, leaving deserts behind. Aeolian activity would subsequently decrease from such areas, while the dust sediment production from the present day main dust sources is likely to have increased during the twentieth and twenty first centuries with glacial retreat (see Fig. 12.3).

(ii) Chemical weathering occurs after the aeolian materials are deposited with some materials being lost (mobile elements). Clay minerals are formed and organic matter added to the soils, with subsequent changes in bulk density. Losses by chemical weathering are not accounted for, nor are the organic matter contents and changes and differences in bulk density (see Arnalds 2010). These methods use bulk thicknesses to calculate rates in mm year^{-1}. Considering these factors and calculating deposition in g m^{-2} accounting for these factors instead of using

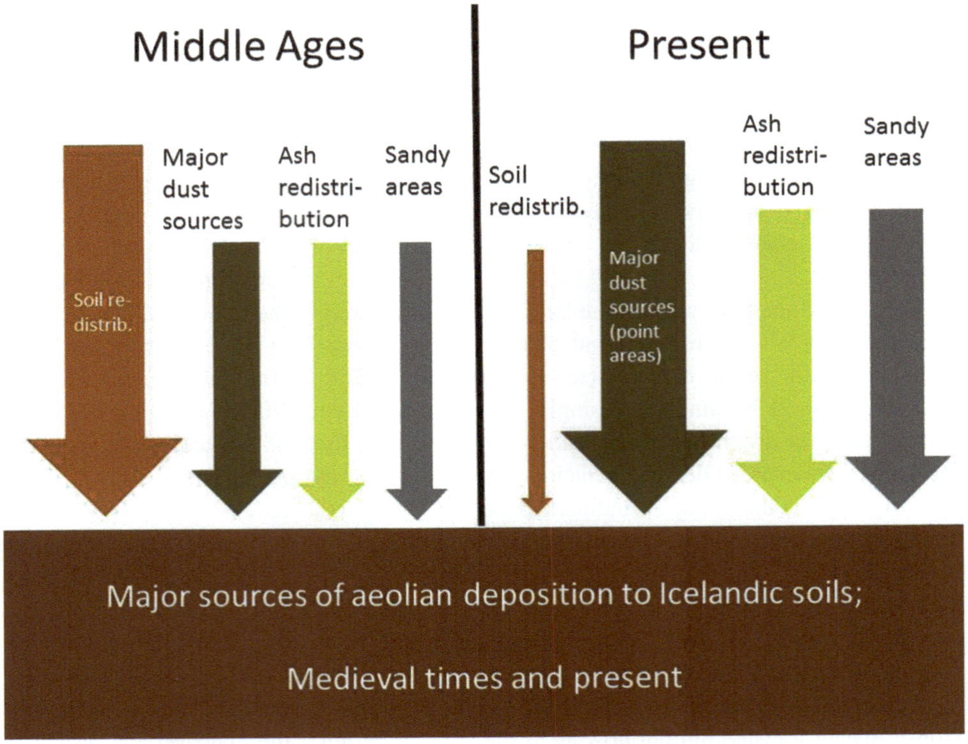

Fig. 12.3 Schematic drawing indicating a shift in contribution from the major dust sources from the Middle Ages to the present. Soil redistribution is presently not as large component compared to earlier erosion periods while major glaciofluvial sources are more active now, together with growing sandy deserts and redistributed ash from the poorly vegetated areas after volcanic eruptions

thickening rates in millimeter, would possibly yield somewhat different results.
Chemical weathering is minimal in areas of high rates, but can be considerable where the rates are low (see Oskarsson et al. 2012 and Chap. 9 on soil genesis). Soil strata found at depths is usually significantly weathered and simply using depth increments between tephra layers would indicate slower accumulation than actual initial rates when these materials were at the surface (materials being lost due to weathering over time). With the weathering, the bulk density of the materials changes from >1.1 to 0.6–0.8 g cm^{-3}, which can lead to underestimation of original aeolian deposition rates.

(iii) Aeolian deposition rates are affected by spikes of aeolian activity after tephra deposition events as exemplified by the 2010 Eyjafjallajökull deposition (Arnalds et al. 2013), but similar large scale spike effects are reported for the 1990 Hudson eruption in Chile (Wilson et al. 2011; Casanova et al. 2013). Volcanic tephra that falls on desert areas, even in small quantities, is unstable, creating such spikes on surrounding vegetation. In Iceland, growing desert areas after the Settlement have possibly amplified aeolian spikes with time.

It should be noted that variable ecosystem surfaces and multiple erosion types make interpretation of aeolian data *per se* not sufficient to understand erosion processes and erosion history at a given site. Wind erosion would not have been the major erosion process in many areas, such as west Iceland with relatively wet soils dominating, and thin dryland soils in between, which are not very susceptible to wind erosion (the area is still relatively well vegetated except for steep hills). Water erosion would be the dominant erosion process, dominating the hillsides. Ideal sites for general interpretations of aeolian records for extensive areas (e.g., highland areas) would be located well within vegetated areas (not at desert margins) relatively far downwind from the degraded areas.

12.2.3 Vegetation—Pollen

Vegetation mapping was started during the middle of the twentieth century, and the results highlighted the poor state of the ecosystems. Vegetation classes in relation to land condition were discussed briefly in Chap. 4, but a system separating land condition classes are introduced later in this chapter. Ingvi Thorsteinsson published a monograph on vegetation protection (*Gróðurvernd*) in 1972 and an overview of the range resources in English (Thorsteinsson et al. 1971). Pollen analyses in Iceland were pioneered by Thorarinsson (1944), Einarsson (1962) and later by Hallsdottir and others (see Hallsdottir and Caseldine 2005), which showed dramatic changes in vegetation composition after the Settlement. Other detailed pollen studies include those of Lawson et al. (2007) from the Lake Mývatn area (NE Iceland) and

Fig. 12.4 Birch trees and vigorous vegetation in the Viðey island (Wooden-Island) in the Þjórsá river in South Iceland. The land in the background is severely degraded, but has been restored to a certain degree. *Photo* © Anna Sigridur Valdimarsdottir

Erlendsson et al. (2009) in southern Iceland, which showed less dramatic vegetation change at the Settlement at the southern coastal location in comparison to many previous studies.

Vegetation studies in areas naturally protected from heavy grazing show vigorous vegetation at such locations, but the studies include those of Jonsdottir (1984) in a protected highland area in North Iceland, and by Valdimarsdottir and Magnusson (2013), who studied an island in the major glacial river of Thjórsá in the South (Fig. 12.4).

12.2.4 Impacts, Resilience and Stability

Volcanic eruptions periodically cause deposition of volcanic tephra (ash). The impact of the tephra is largely determined by the thickness of the tephra that is deposited and vegetation height receiving the deposition (Arnalds 2013). Composition of the vegetation is also important, with mosses and lichens being vulnerable to deposition impacts. Grazing affects vegetation height, composition and nutrient levels in the soil, and long-term heavy grazing reduces both resilience and stability of the systems. The relationship between ecosystem resilience and stability on one hand and the periodic impacts of volcanic events and cold spells on the other has been stressed by many (see Arnalds 2013). The time of dramatic ecosystem changes begins immediately after the Settlement around 870, but exact timing of disturbances varies somewhat. This is reflected in multidisciplinary studies (e.g., Gudbergsson 1996; Geirsdottir et al. 2009; Vickers et al. 2011), that show that time of initial major disturbance is affected by such factors as climate (mainly cold spells), landscape, elevation, land accessibility (and cost of woodland clearing), and ecosystem stability and resilience, which in turn are affected by soil texture, soil thickness, and the presence of coarse tephra layers. It has been noted that many of the marginal areas such as highland areas were subjected severe land degradation with widespread soil erosion early on (Gudbergsson 1996, 1975; Gudmundsson 1997). Degradation was delayed in many areas of higher resilience such as in the West.

It is worth noting that climate had become cooler leading up to the Settlement, with decline in birch pollen (see e.g., Lawson et al. 2007), making the systems more vulnerable to anthropogenic disturbance. Olafsdottir et al. (2001) concluded that the forest cover and vegetation was already declining at the time of the Settlement because of a cooling trend after 3000 BP that culminated between about 1300 and 1900. The cold Middle Ages were, however, characterized by "broad interval of warmth" with the occurrence of "multidecadal cold intervals," as was determined by investigation of lake sediments (Geirsdottir et al. 2009). The natural systems with limited anthropogenic influence, can adapt to such climate changes and most of the systems would have ample resilience and stability to counter cold spells, with warmer periods in between to enhance recovery. Cooling during the Middle Ages would, however, affect the elevation of the upper limit of continuous vegetation, which became gradually lower (Olafsdottir et al. 2001). The cooling trend would have reduced the stability of the ecosystems with detrimental consequences when human land use increased, especially at marginal locations.

Fig. 12.5 Fate of the woodlands. Continuous heavy grazing by sheep prevents regeneration of the birch woodlands, eventually leading to their disappearance

Grazing in Iceland involved very damaging land use practices such as winter and early spring grazing, which are particularly harmful for declining and marginal ecosystems with low stability and resilience (Fig. 12.5). It has been argued that the grazing capacity of the land had not been exceeded during the Middle Ages judging from total medieval livestock numbers. Such calculations, based on average values for climatic conditions, animal numbers, and vegetation are rather meaningless, as they do not consider ecological principles, including stability, resilience, fluctuating climate, and periodic volcanic impacts. Poor grazing practices were indeed very detrimental: grazing during cold spells and after tephra deposition events is more harmful than grazing during average years in relation to impacts on ecosystems.

The frequent volcanic eruptions have been very destructive for Icelandic ecosystems, with less stable and resilient systems after the Settlement. A good example is found in the Askja area. During the nineteenth century, a growing population led to the establishment of new farms in many marginal areas, such as in the eastern highlands (Jökuldalsheiði). A volcanic eruption in the Askja volcanic system in 1875 lead to the deposition of rhyolitic tephra over the area, with severe erosion following and subsequent land abandonment (Thorarinsson 1979). Volcanic eruptions can also reduce the stability and resilience of systems to repeated volcanic activity, but Dugmore et al. (2007) noted such repeated volcanism as an important factor for the massive destruction in Þjórsárdalur, close to Mt. Hekla, caused by the 1104 and 1300 AD eruptions in the volcano. The largest Hekla eruptions, depositing thick rhyolitic tephra over extensive highland areas, are bound to have caused severe ecosystem disturbance, even during prehistoric times with no land use (Fig. 12.6).

Fig. 12.6 Thick, coarse, rhyolitic Hekla layers in soil (*yellowish*). These layer lacks cohesion and are very susceptible to wind erosion if exposed, causing damage to surrounding systems after erosion processes are initiated. Grains >2 mm in diameter are common and are easily moved by the wind

Fig. 12.7 Unstable ecosystems within the volcanic belt. Much of this area has become a barren desert, with ecosystem remnants surviving in landscape positions providing more resilience (more moisture etc.)

This has been shown to be the case from studies of lake deposits in Hvítárvatn (South Central Iceland) by Larsen et al. (2011), where severe erosion continued for over 100 years after the deposition of the H3 tephra (probably the largest tephra deposition over Iceland during Holocene), dated at about 3100 BP by the authors. However, the land eventually recovered, even in highland areas where the tephra was >10 cm thick. Reduced land use in many areas at present has resulted in evidence of reduced soil erosion. Reduced land use occurred also during the Middle Ages, when the population decreased dramatically because of plagues. Streeter et al. (2012) showed that reduced land use following population decline in the fifteenth century resulted in reduced "geomorphological activity" such as soil erosion.

Climatic variations, mainly cold spells, are often stressed as important contributors to ecosystem collapse in Iceland. Yet, the severity of the "Little Ice Age" is debated (Ogilvie 2005), although the occurrence of cold spells during this time are not disputed. The impact of cold spells would, however, largely depend on the stability and resilience of the systems. Systems under heavy land use would be more susceptible to climatic and volcanic impacts, and the destruction of the woodlands is bound to have an immense impact on the stability of the systems with a change in microclimate. Heavy land use may also mask the impacts of cold spells judging from lake sediments in the Westfjords (Doner 2003), especially in places where the aeolian soil mantle is relatively thin as in that area. Streeter et al. (2012) noted delayed impacts of climate change, attributed to landscape or ecosystem resilience. But one has also to bear in mind that colder climate also means less biological production, which results in increased grazing pressure if animal numbers remain similar, with subsequent reduced food production and consequently this interaction can result in the well-established downward spiral of ecosystem destruction (e.g., Whisenant 1999). The coldest years also saw extensive sea ice at the northern coastlines, blocking fishing harbors and limiting fishing options (see Ogilive and Jonsson 2001) and still increasing the pressure on land-based resources. These authors noted that land use impacts were likely to dwarf the impacts of potential climate change such as the effect of the "Little Ice Age." The question of climate change versus land use effects is very important globally today (see Herrick et al. 2013), with some arguments stating that land use may, in many cases, dwarf the effects of present day man-induced climate change.

The stability of ecosystems varies considerably between geographic regions in Iceland (Figs. 12.7 and 12.8), with wetland soils and more shallow dryland soils far from major aeolian and volcanic sources being more stable than dry thick soils with coarse layers (tephra and aeolian layers). Geirsdottir et al. (2009) noted that erosion had begun around lake Haukadalsvatn in West Iceland prior to the Settlement due to cooling climate, with no sediment spike deposited into the lake soon after the Settlement. This area is characterized by more stable ecosystems than the volcanic zone, with limited wind erosion, and it is still an area with relatively continuous vegetation cover.

Fig. 12.8 Relatively stable and resilient ecosystems in East Iceland, especially the wetlands in the valley. The underlying rock is Tertiary basalts with low permeability leading to high water tables in the area. The dryland positions are evidently more unstable

12.3 Soil Erosion

Soil erosion has long been considered the major environmental problem of Iceland. This prompted a major effort to map soil erosion in the country, which was carried out between 1991 and 1997. The aim of the project was to obtain an overview of soil erosion in all of Iceland, but the extent of the problem was much disputed at that time. The project was run by the Agricultural Research Institute and the Soil Conservation Service of Iceland. The scale of mapping was 1:100,000. Mapping was carried out in the field, with polygons drawn on Landsat five satellite images, and the data was subsequently stored and handled using GIS software. The results were published in the book *Soil Erosion in Iceland* (Arnalds et al. 1997) which was translated into English (Arnalds et al. 2001). Various efforts were made to inform land owners and the general public about the state of the land in relation to the erosion mapping project, including the booklet *To read the land* (Arnalds 1997). The project received *the Nordic Nature and Environmental Award* in 1998. This chapter outlines the methods and results of the erosion mapping in Iceland.

12.3.1 Erosion Forms

Soil erosion processes in Iceland are extremely varied and many processes can occur at any given site. Snow cover and the action of frost are important factors influencing the erosion, including needle ice formation (see Chap. 10). Conventional models for soil erosion, such as the Universal Soil Loss Equation (USLE) and the Wind Erosion Equation and similar models are poorly suited for Icelandic conditions except perhaps for the desert areas. Conventional methods are largely designed for cultivated areas and are often poorly suited for rangelands in general (e.g., SRM 1992; NRC 1993; Pierson 2000). A geomorphological approach was taken for designing methods for Icelandic conditions. Soil erosion forms were defined based on features that can be identified in the field. These forms are broadly divided into two categories: (i) forms associated with erosion of soils with vegetation cover (Andosols), and (ii) erosion on deserts, Vitrisols (vitric soils of the deserts, see Sect. 6.5), Leptosols etc. (Table 12.1).

Table 12.1 Classification of Icelandic erosion forms (Arnalds et al. 2001)

Erosion forms associated with erosion of vegetated land	Erosion associated with desert landscapes
Advancing sand fronts	"Melur" (lag gravel /till surfaces)
"Rofabards" (escarpments)	Lava surfaces (without sand or soils)
Erosion spots	Scree (gravelly/rocky hillsides)
Erosion spots on solifluction slopes	Sandur (aeolian and tephra sand surfaces)
Landslides	Sandy lag gravel
Water channels (rills and gullies)	Sandy lavas

The Icelandic soil erosion survey system includes grading of erosion severity, with a scale of 0–5 as follows: 0: no erosion; 1: little erosion; 2: slight erosion; 3: considerable erosion; 4: severe erosion; and 5: very severe erosion. Areas with erosion severity 4 and 5 are not considered suitable for grazing until erosion has been halted. Following is a discussion on these different erosion forms with examples illustrated. Examples of the erosion severity grading are also illustrated with some of the photographs.

12.3.2 Advancing Sand Fronts (Encroaching Sand)

Advancing sand fronts form where there is a source of loose silty and sandy sediments that are carried by the wind over vegetated areas. A tongue-shaped front into the vegetation cover is created, where sand is advancing (Fig. 12.9). The soil materials underneath the destroyed vegetation are added to the silty/sandy materials that keep advancing with prevailing dry winds. A barren desert is left behind. If the soils are very sandy, e.g., if they have thick coarse tephra layers, the soil materials added to the process along the way of the front are sufficient to maintain the process; the tongue continues until the advancement is stopped by some landscape features such as a hill, river, or a shoreline. Continuous activity with advancement of the front is also maintained if there is a steady source of aeolian materials, such as are deposited by repeated flooding of glacial rivers or from glacial lakes with unstable shore lines. Advancing sand fronts are common on and near the volcanic active zone, and the remnants are expressed as long straight lines on the landscape on the boundary between vegetation and desert, in the same direction as the dry winds, from the abrasion of the silt/sand along the vegetation–desert boundary.

Fig. 12.9 Example of the formation of an advancing sand front. The original source of the sand (and silt) is sediments half filling the lake due to water erosion in the surrounding area, drying out the lake. The sediments are driven to the south (*left* on the fig. 12.9) with north prevailing dry winds

There is evidence that this type of erosion has been the most destructive erosion form after the Settlement. Advancing sand fronts can progress over vegetated areas up to several hundreds of meters each year if there are a lot of sandy and silty materials advancing. There are accounts of such rapid advancement in Northeast Iceland, where the sediment sources vary from river sediments (Jökulsá á Fjöllum, see Fig. 12.10), to unstable ephra and glacial deposits.

The Icelandic Soil Conservation Service was established as *Sandgræðsla ríkisins* in 1907, mainly to battle advancing sand, which was destroying farm after farm in South Iceland. An account of the battle against the sand is given in several papers in a book commemorating the 50 years anniversary of the Soil Conservation (Sigurjonsson 1958; in Icelandic), but the history is also accounted for in the 100 years story of soil conservation in Iceland (Olgeirsson 2007; in Icelandic) which in part was used as the basis for an book in English by Crofts (2011): *Healing the Land.*

Areas that are subjected to sand abrasion from sand sources, or possibly subjected to such stress, like the areas within the volcanic zone, can be described as supermarginal areas, where it is very important to maintain as strong vegetation cover as possible, preferably birch or salix woodlands/shrublands, which alter wind conditions at the surface and accumulate and store snow cover. Many of these areas are now protected from grazing, but recovery of sandy deserts is very slow because of the unstable sandy surfaces (see subchapter on restoration later on).

12.3.3 Rofabards

"Rofabards" may be the most typical of all the Icelandic soil erosion features and most noted by observers. Rofabards are escarpments marking the boundary between vegetated systems on top with a mantle of Andosols formed in aeolian and tephra materials, but a barren surface at the bottom and in the adjacent area; typically till, lava or a sandy surface. Rofabards are often very prominent features on the landscape (Fig. 12.11) and were the subject of discussions by Arnalds (2000). The vegetation and the root mat provides a cohesive upper surface, but underneath are less cohesive soil materials which are undermined by erosion processes (Fig. 12.12). Conditions for the formation of rofabards are found in dryland areas where the aeolian soil mantle is relatively thick, which is within and near the active volcanic zone and close to glaciers. The thicker and coarser this aeolian soil mantle is above the harder bedrock, the more active the rofabards become in terms of erosion.

Fig. 12.10 Infrared satellite image of a portion of Northeast Iceland showing the path of many advancing sand fronts. Vegetation is *red* on the image, but the deserts *dark-bluish*. Lake Mývatn is at the *bottom left*. Note the *straight lines* on the sides of the path of the fronts. New lava (1980s) from the Krafla system shows up *black* on the *center left*. The river in the *upper right corner* is Jökulsá á Fjöllum, the major river of NE Iceland. Eruptions in volcanic systems under Vatnajökull ice cap causes catastrophic flooding which deposits large amount of sediments and which are later subjected to wind erosion and the creation of advancing sand fronts. The Hólsfjöll desert was created by such process during the Middle Ages, the formation aided by reduced stability of the systems due to land use. A desert area NW of Lake Mývatn originates in sandy deposits near the lake. Numerous smaller advancing sand fronts are east (*right*) and north of Lake Mývatn. SPOT 5 Satellite image © EruoImage

Measurements show that the rofabards are retreating (the boundary between desert and vegetated areas are moving into the vegetated land) at rates ranging from few millimeters to >50 cm year^{-1}. It should be noted, however, that much of the most erosion susceptible soils have already been lost, and rates of soil loss were rapid during the main erosion phases of the Middle Ages.

Many soil erosion processes occur at rofabards. Wind erosion is common, especially in NE Iceland and where the soils are thick and coarse. Water erosion is important for

Fig. 12.11 Rofabards are often quite prominent in the landscape within the volcanic zone and close to active aeolian sources (creates thick Andosol mantle), but they were more abundant in older days during the main erosion phase of the Middle Ages

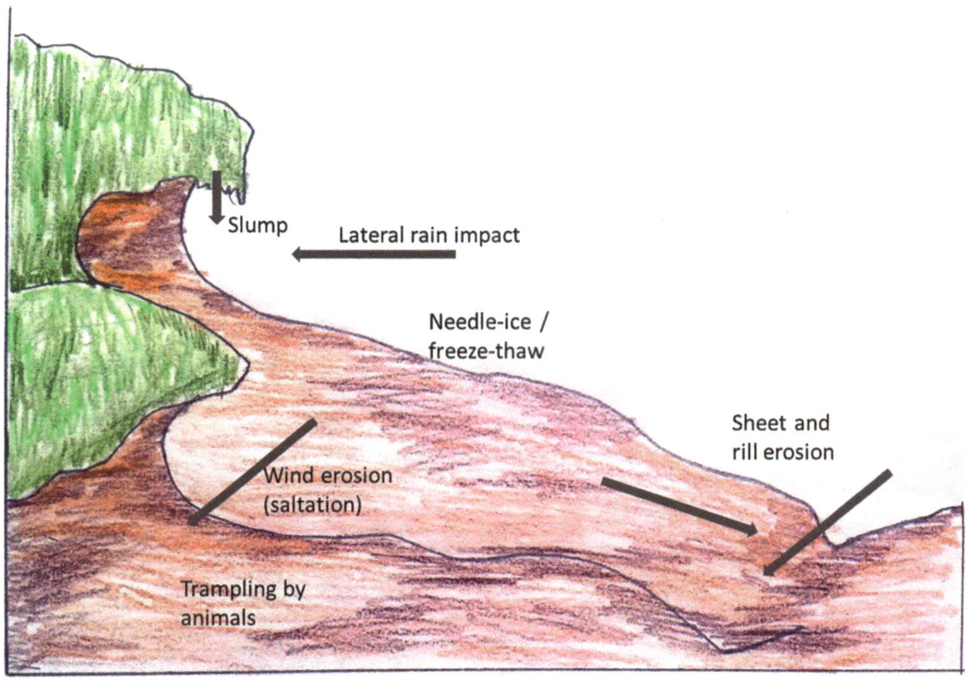

Fig. 12.12 Erosion processes that can be expected at rofabards

rofabards in South Iceland. Lateral rain, during high intensity winds is a key factor contributing to the erosion of rofabards as well as to erosion of desert surfaces (see Figs. 12.12, and 12.13). As the rofabards become undermined, the surface turf mat slumps down. Sheep seek shelter under the rofabards, and trampling by the animals also contributes to the erosion.

12.3.4 Erosion Spots

Erosion spots are small patches of barren soils in otherwise vegetated surface (Fig. 12.14) which typically is relatively flat. They are extremely common in the Icelandic rangelands, and are usually the result of overgrazing, at least when they occur in the lowlands. Vegetation re-establishment is

Fig. 12.13 Lateral rain splash during one high intensity storm has cut into the ground but small pedestals are protected by the pebbles. Large rain drops driven by 20–40 m s^{-1} gale force winds have a substantial erosion power. Grass straws at *top* and the *bottom right* give the scale. *Photo* © Elin Fjola Thorarinsdottir

Fig. 12.14 Erosion spot are barren patches in otherwise vegetated land. They are most often a result of overgrazing, especially in the lowlands. The vegetation here is degraded poor heathland and land covered with moss, with low abundance of grazing plants, which have been grazed out over the years

slow after the erosion spots have formed. Conditions for needle-ice formation are optimal within the spots, which impede vegetation establishment in the spots by killing little seedlings that get established during summer. Recovery, after prevention from grazing, can take many decades.

12.3.5 Hill Slopes: Solifluction, Water Channels, and Landslides

Hill slopes are subjected to slow downhill movement of the soils, which is termed "solifluction". Solifluction leads to the formation of wave-like system of small terraces ("steps") or

Fig. 12.15 A front of a solifluction lobe with erosion spot which is caused by overgrazing by horses. Erosion can develop into a severe problem on such slopes

tongue-shaped lobes (Fig. 12.15). Erosion spots that form on such slopes are subjected to the force of running water, which is a minimal factor for erosion spots on flat land. Therefore, they are a separate entity in the Icelandic soil erosion classification system. Erosion spots on hill slopes are often easier to see looking down the slopes, and can be pronounced even though the hill looks full vegetated looking up the hill (Fig. 12.16).

Water channels occur on many vegetated hill slopes, but often develop to form rofabards as the erosion progresses on slopes with thick aeolian soil mantle. Hill slopes are most common in the valleys of the Tertiary lava stack landscapes of Iceland. Slope length is often limited by each lava flow in the stack, making water erosion not as serious as it would be if the slopes had continuous surfaces. Palagonite formations (hyaloclastites from Quaternary volcanism, see chapter on geology) tend to have long continuous convex slopes. They are periodically subjected to intense water flow and are frequently lacking vegetation cover, but may often have had

Fig. 12.16 Looking down solifluction slopes reveals the erosion spots, which are not prominent when the view is from *below*. This slope looks fully vegetated looking up the slope

Fig. 12.17 Soil and vegetation remnants on otherwise barren scree slopes in the Westfjords. Many of the barren hill slopes of Iceland were in part vegetated prior to the Settlement. Vegetation recovery on these kinds of slopes is extremely slow, especially on rhyolitic scree slopes

such cover prior to use of the land. Such vegetation and soil cover may have been lost early on after the Settlement, but remnants of these ecosystems can be found on many of the eroded hillsides (Fig. 12.17).

Landslides are common in Iceland and they sometimes cause damage to buildings, roads, and other structures. The reason for the high frequency of landslides is the noncohesive Andosols and periodic overloading of water. The soils can adsorb very large quantities of water and some become thixotropic upon reaching the liquid limit. Large mudslides can occur where conditions lead to saturation of the soils, such as during snow melt in spring (Fig. 12.18). It seems that both tephra layers in the soils and the bedrock underlying the aeolian soil mantle can provide slip planes that further increase landslide susceptibility.

Fig. 12.18 A large landslide in N Iceland, formed by oversaturation of water uphill and the soil reaching the liquid limit

12.3.6 Erosion Associated with Desert Landforms

The Icelandic desert systems cover over 40 % of the island and they are subjected to erosion by wind, water, hill slope process, and cryoturbation. Desert ecosystems were discussed in Chap. 4. Soil classification of desert soils was discussed in Chap. 6, and soil properties (Vitrisols) in Chaps. 7–9. The surface types of Icelandic deserts were classified and mapped during the "Soil Erosion Mapping Project" that provides the basis for this subchapter (Arnalds et al. 1997, 2001). As was stated previously, the Icelandic erosion classification considers soil or surface stability with scores from 0 to 5 (no erosion—extremely severe erosion). The desert types are **lag gravel surfaces** (glacial deposits dominate), **sandy surfaces** (fluvial and aeolian deposits dominate), **barren lava surfaces,** and **scree slopes**. As aeolian processes carrying silt and sand over various surfaces are such a dominant feature of Icelandic nature, subclasses with **sandy lag gravel** and **sandy lavas** are special classes in this classification (see Chap. 11 on the aeolian

Table 12.2 Desert surfaces in Iceland

Desert surface	Erosion score					Total
	1	2	3	4	5	
	km^2					
Lag gravel	9,939	8,546	6,580	0	0	25,065
Lavas	1,832	228	25	0	0	2,085
Sandur	195	337	318	1,087	2,828	4,765
Sandy lag gravel	8	741	5,407	6,217	1,286	13,659
Sandy lavas	10	101	1,366	1,757	1,620	4,855
Scree slopes	64	913	2,378	1,255	392	5,002
Bare brown soil	17	518	350	65	36	987
Total	12,065	11,384	16,424	10,381	6,162	56,418

More than one desert type can occur within the same mapped polygon; hence the total area in the table is considerably higher than the actual value

environment). The results of mapping of the deserts are presented in Table 12.2 and Fig. 12.19.

Lag gravel surfaces are the most common desert surfaces, extending about 25,000 km^2, but much of these surfaces occur in combination with vegetated land, hence the nearly 10,000 km^2 with erosion score 1. The lava surfaces are also rather stable unless sandy deposits have been blown onto such surfaces. The sandy surfaces have high erosion scores (very unstable surfaces), and they occupy >20,000 km^2, as the surface types sandur (sand-flats), sandy lag gravel and sandy

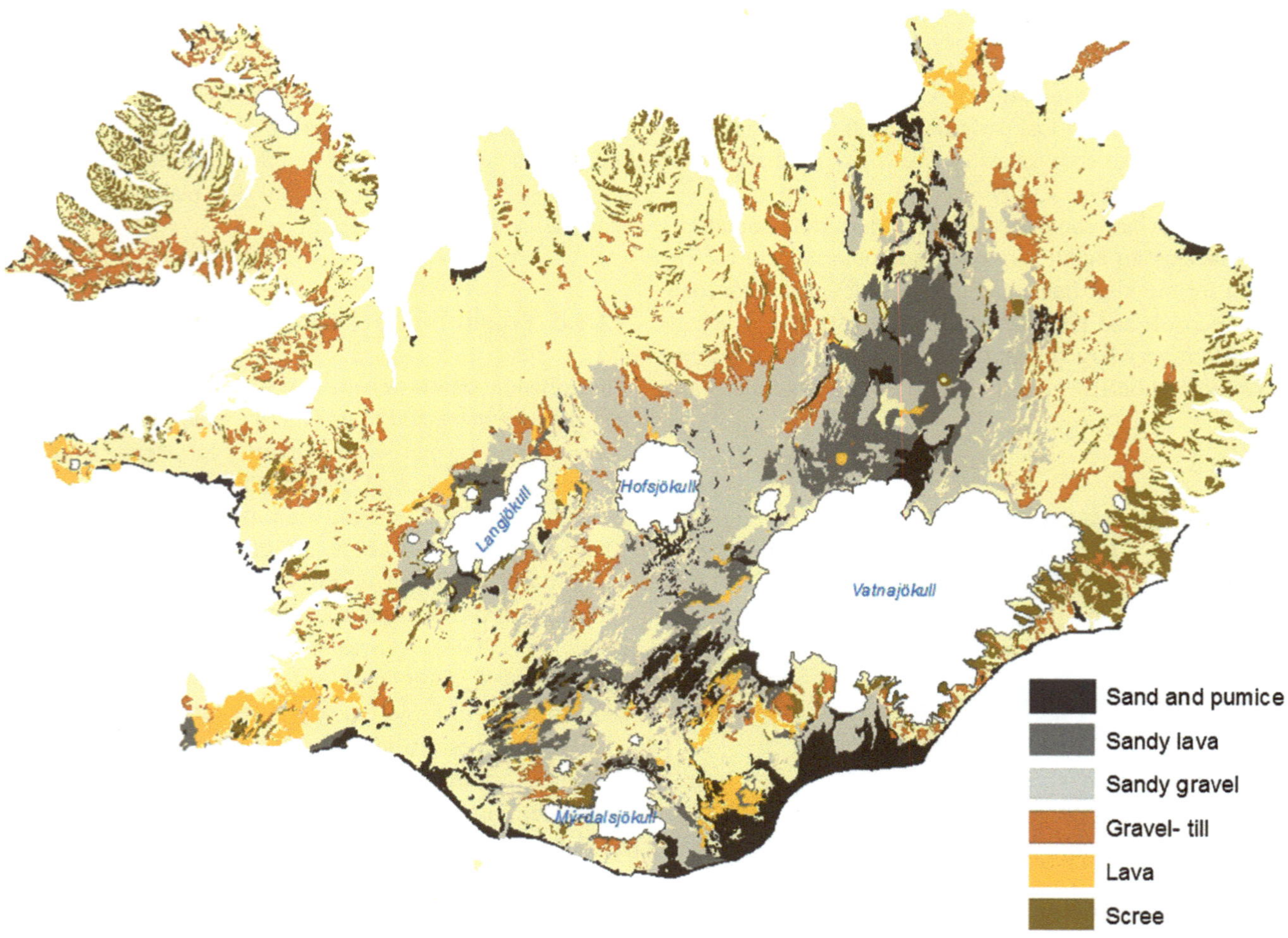

Fig. 12.19 Map showing the extent of desert surfaces in Iceland. Modified from Soil Erosion in Iceland (Arnalds et al. 2001), prepared by Sigmundur Helgi Brink, © OA/SHB, AUI

lava areas. A large proportion of Holocene lavas have sandy deposits in the surface (4,855 km^2). Scree slopes are quite common with >5,000 km^2. It should be noted that each unit or polygon under this mapping scheme can have several erosion scores (e.g., both sandy lag gravel and sandur occurs within the polygon), resulting in a total of 56,418 km^2 area mapped as desert, but the total area of deserts is about 40,000 km^2 according to the Nytjaland database (see Chap. 4).

Most of the desert areas continue to be grazed, in spite of the negative impact of such land use practice on these collapsed and/or vulnerable ecosystems.

12.4 Wetland Disturbance

For ages, wetlands provided grazing areas, they were important for hay production and some provided peat for fuel and turf for house structures (Magnusson 1998). Yet, wetlands also were obstacles to the traveler and they were, in a sense, considered a negative aspect of the landscape in older times. Draining the wetlands facilitates oxidation of the organic matter and release of nutrients, resulting in substantial increase in biomass production (Magnusson 1998), a prospect that was of interest to the farming community during the early twentieth century.

After World War II, with the arrival of large and effective digging machinery, a large scale program to drain the wetlands was initiated (Fig. 12.20). This draining program was a part of the national post-war strategy to ensure food safety (Gardarsson et al. 2006). In the early stages of the drainage effort (1940–1965), the drained areas were mainly used to establish permanent hay fields for the production of winter fodder for cattle and sheep (see Helgadottir et al. 2013). During later stages of the effort, the focus was on creation of rangeland. Currently, less than 15 % of the drained areas are used for agricultural purposes as hay fields or for growing grain. The Agricultural University of Iceland (unpublished data) has mapped and measured the length of the drainage ditches to be >30,000 km, with >3,500 km^2 wetland area converted to grassland (Hallsdottir et al. 2013). It has been estimated that 50–75 % of all wetlands in lowland areas have been altered and disturbed (Oskarsson 1998; see also Thorhallsdottir et al. 1998) (Fig. 12.21).

The drainage of these wetlands has resulted in large scale greenhouse gas emissions as the organic matter oxidizes; a release of greenhouse gases of the same order of magnitude as emissions by the heavy industry (smelters) in Iceland (Oskarsson 2008; Gislason 2012; Hallsdottir et al. 2013). The majority of the drained wetlands are not used for hay production and some even not for grazing. It can be inferred by the published literature (Geirsson 1998; Oskarsson 1998; Magnusson 1998; Wald 2012; Olafsdottir 2013) that some were drained simply because of generous financial aid provided for the effort, especially after 1974, and that the vast majority of wetlands drained after 1966 are generally not in use as hay fields.

Fig. 12.20 Drainage ditch being put in a wetland in the South. The ecosystem changes towards a grassland, but a large amount of CO_2 is released by oxidation of the organic rich wetlands. *Photo* © Hlynur Oskarsson, AUI

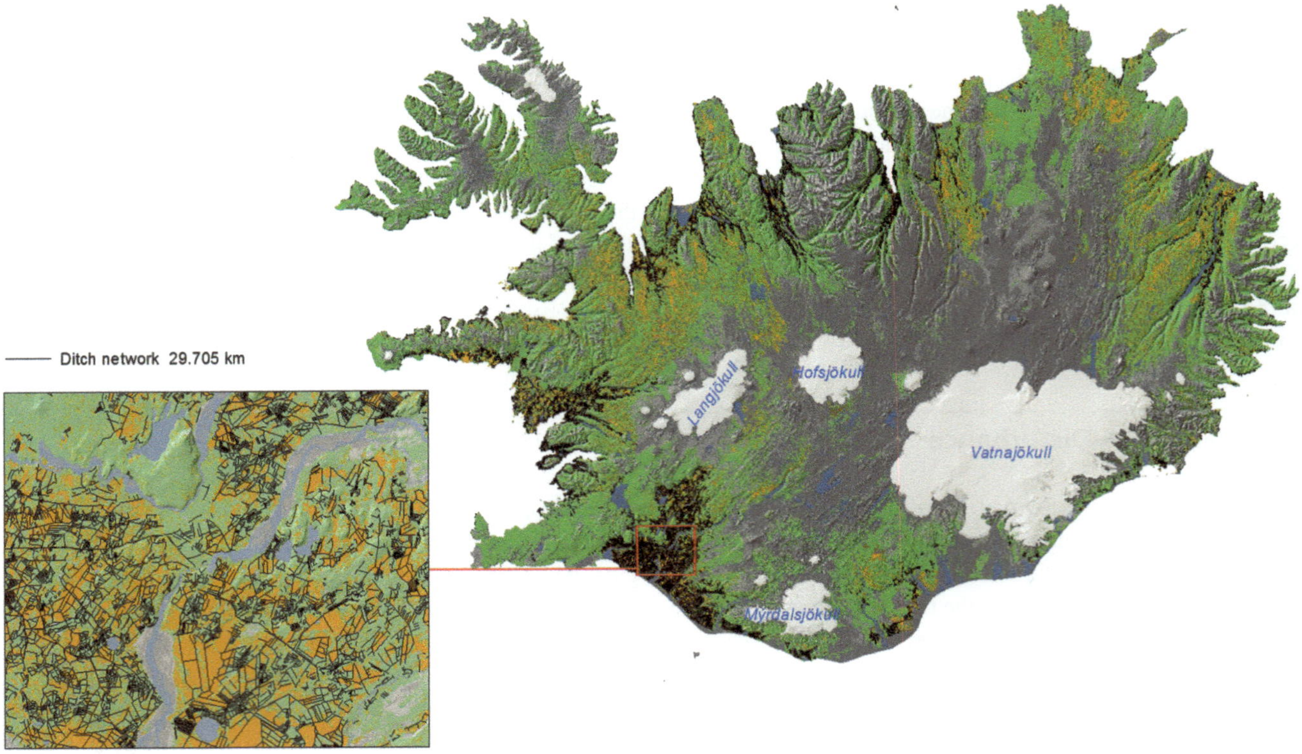

Fig. 12.21 Wetland disturbance by drainage. The wetland areas are shown in *red* and *pink colors*, but the drainage ditch system has been superimposed as *thin black line* following each ditch. The figure clearly indicates that much of the Icelandic lowland wetlands have been disturbed. A portion of the map is enlarged, to give an indication of how ditches dominate many of the wetland landscapes. *Image* Icelandic Farmland Database (vegetation classes) and the IGLUD database (drainage ditches), prepared by Sigmundur Helgi Brink. © SHB/JG/OA, AUI

Wetland restoration has begun in Iceland and will likely grow in importance during the next decades. However, some of the drained wetlands are still very important for agriculture and will in all likelihood continue to be used as such.

12.5 Reading the Land—A Simple Scheme for Land Condition

12.5.1 A Simple Land Condition Scheme

The observer of Icelandic landscapes, native or foreign, may believe that the beautiful Icelandic landscapes are pristine and unspoiled by human actions. The observer may not comprehend the vast changes that Icelandic landscapes have undergone since the Settlement of the country. Some efforts have been made to prepare materials to aid the public in understanding the state of Icelandic ecosystems. Arnalds (1997) published an educational booklet titled *To Read the Land* and Arnalds and Aradottir (2014) recently published a book in Icelandic (*To Read and Heal the Land*), where an attempt is made to explain for the public ecosystem disturbance and the condition of the land. These publications draw from many sources, including the "Erosion Mapping Project" and a paper published in the Soil Conservation Yearbook on assessment of land condition in Iceland (Aradottir et al. 1992). The methods presented by Aradottir et al. (1992) which are used here were in part adopted and modified by Archer and Stokes (2000) and Bestelmeyer et al. (2011) for rangelands in general and by Thorsson (2008) for an Icelandic desertification research study. Following is a short discussion on the classification land condition in Iceland, based on the original Aradottir et al. (1992) system to aid the reader in understanding the Icelandic landscapes.

Aradottir et al. (1992) split land condition into six stages, ranging from pristine (Stage I) to deserts (Stage VI) (Fig. 12.22). There are dramatic differences from the pristine stage to the desert for most ecological variables. Table 12.3 lists some of the key surface and soil parameters with grading (A–E) of surface parameters and a numerical range for three soil parameters to give an indication of the dramatic changes in ecosystem functions. The discussion below is mainly aimed for dryland systems below 300–400 m elevation.

I. **The pristine stage** is characterized by woodlands and rich heathlands with grasses and herbs (Fig. 12.23). The soils are Brown Andosols with fertile soils.

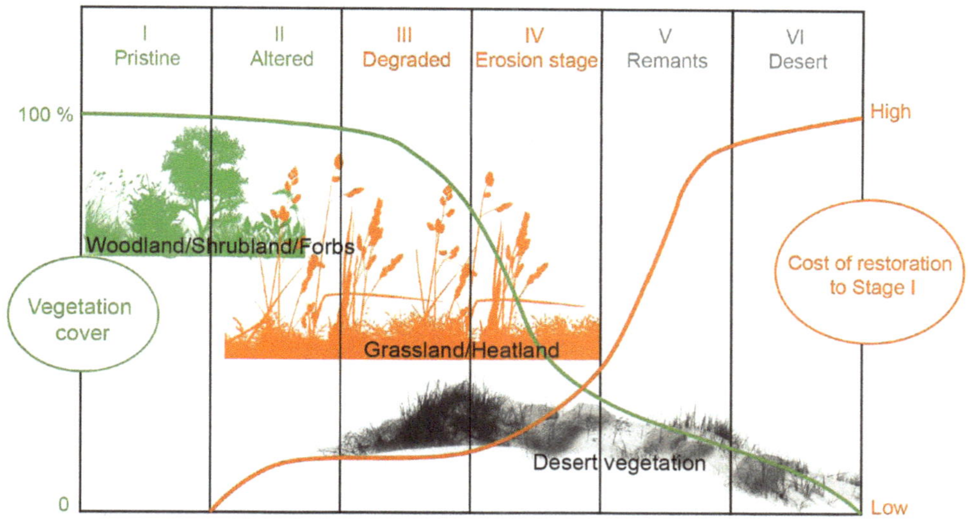

Fig. 12.22 Land condition classes, vegetation cover and cost of ecosystem restoration to Stage I. Adopted from Aradottir et al. (1992) and Arnalds and Aradottir (2014)

Table 12.3 A gradient for dryland condition for factors typically used applied to assess land condition or quality

	I. Pristine	II. Altered	III. Degraded	IV. Erosion	V. Remnants	VI. Desert
Surface						
Plant cover (%)	A full	B	C	D–E	E	E non
Surface wind	A non	B	D	E	E	E extreme
Litter	A high	B	C	D	E	E non
Snow accumulation	A high	B	B–D	D–E	E	E non
Wind erosion	A non	A–C	B–D	C–E	C–E	C–E extreme
Water erosion	A non	A	B–C	C–E	C–E	C–E extreme
Needle ice formation	A non	A	B	C–D	E	E extreme
T° extremes	A non	B	B–D	C–E	E	E extreme
Evaporation	A limited	B	C	D–E	E	E extreme
Infiltration winter	A ample	B	C–D	D–E	D–E	E limited
Topsoil						
C% in top 30 cm	6–12	3–7	1.5–4	0.5–3	0.2–2	0.0–0.5
N top 30 cm, kg/ha	>10,000	>5,000	>2,500	<1,000	<500	<250
Clay%	15–30	10–25	10–20	3–15	3–10	1–5

Descriptive terms (non, low, considerably, high, very high) or typical numerical values or ranges. Note that moss and poor heathland may also occur as successional stages, e.g. on disturbed ground (lava, ash, floods) or as a result of restoration efforts

Erosion is negligible, the system has favorable microclimate with limited surface wind speeds and snow is trapped in winter for insulation and better moisture conditions. Infiltration is considerable in winter. Carbon contents are high, exceeding 6 % in surface horizons with ample nitrogen stocks and robust nutrient cycling.

II. **The altered stage** is characterized by systems that have been modified by land use, without causing severe land degradation (Fig. 12.24). The shelter from the woodland is lost, allowing for more effects of cryoturbation with increased formation of hummocks. Productivity is lower with less amount of nutrients (carbon, nitrogen) in the system compared to Stage I. Erosion is still low.

III. **The degraded stage** is often characterized by hummocky landscape (Fig. 12.25) with systems with high proportion of mosses and poor heathlands (see Chap. 4 on vegetation classes). Microclimate is unfavorable with intense cryoturbation. Litter accumulation is limited and physical surface characteristics are notably more unfavorable than at Stages I and II, with reduced infiltration in winter and surface runoff during snowmelt events. There is more erosion activity and more coarse materials being deposited on top of the soils, leading to lower C, N, and clay contents in surface horizons.

Fig. 12.23 Stage I. Pristine. A typical woodland provides good shelter, and robust nutrient and water cycles. Soils are rich in organic carbon with favorable soil properties and good infiltration the year around

Fig. 12.24 Stage II. Altered and somewhat degraded stage. The carbon pool is lower with less robust nutrient cycles. Little shelter from the wind and there is pronounced cryoturbation with thufur formation. Water is lost with blowing snow and with impeded infiltration in winter. Some erosion spots visible on the landscape. Species composition has been altered by selective grazing, with less palatable species becoming more abundant than grasses and forbs

IV. **The erosion stage** is characterized by a mosaic of two systems: vegetated areas and barren patches where erosion has removed the aeolian soil mantle (Fig. 12.26). The vegetated areas have the characteristics of degraded areas (Stage III) and the barren areas have the features of desert surfaces. Erosion is quite rapid at this stage.

V. **The remnant stage** is characterized by the deserts of Stage VI, yet there are remnants of previous ecosystems (Fig. 12.27). These remnants bear similarities to systems at Stage III (degraded), but with lower nutrient reserves due to aeolian and fluvial deposition of coarse materials. The remnants are often important sources of

Fig. 12.25 Stage III. The degraded stage. Considerable amount of soils has been lost together with vegetated systems, with barren patches becoming prominent on the landscape. Active erosion. The soil properties of the barren areas are poor compared to the vegetated patches. The area shows some signs of improvement with reduced grazing intensity with relatively vigorous plant growth on the vegetated patches

Fig. 12.26 Stage IV. The erosion stage (late in the stage); soils and vegetation are lost at a rapid rate. The rofabards show signs of high erosion activity. The soils here are coarse grained and especially susceptible to wind erosion

plant materials to spread around the surrounding desert during restoration efforts.

VI. **The desert stage** is the final stage of the degradation (Fig. 12.28). The soils are Vitrisols (Sect. 6.5) or Leptosols. Plant cover is often 5–15 %. Here is no shelter from the wind, the desert surface heats up during summer leading to very rapid evaporation, but there is very limited infiltration of water during winter. This leads to intense surface runoff during high intensity snowmelt events in winter. In summer, the sandy areas are subjected to intense wind erosion events (see Chap. 11 on aeolian environments). The soils are coarse with limited nutrient reserves and low water holding capacity compared to the vegetated systems.

Fig. 12.27 Stage V. The remnant or vegetation island stage. Vegetation and soil remnants indicate lost ecosystem riches. Deserts dominate the landscape, with lack of energy input (photosynthesis), poor nutrient cycling (soil biota, carbon, and nitrogen) and disrupted water cycle. Desert surfaces are shaped by the geology of each area, such as lavas, sandy surfaces, scree slopes, and gravel. The islands are important sources of plant materials for spreading during restoration efforts

Fig. 12.28 Stage VI. The desert stage. All the previous soils and vegetation cover has been lost. Vitrisols, poor in nutrients and with disrupted water cycle remain (see Table 12.3 for changes in soil and surface parameters). Extreme surface conditions with excessive evaporation in summer but concrete ice and limited infiltration in winter

12.5.2 Condition of Communal Grazing Areas

The Erosion Mapping Project resulted in an assessment of all Icelandic communal grazing areas (Arnalds et al. 1997, 2001). The grazing commons are typically 400–1,000 km^2. There is a large difference in condition between the commons. Some have reasonably good vegetation cover (>80 %), often characterized by poor heathland and sometimes wetlands. The communal areas rated in poor condition often have less than 50 % vegetation cover and severe erosion (erosion grades 4 and 5) on >50 % of the land. Yet, almost all these areas are still used for grazing, regardless of their condition. Most of these poor-state commons are within or in the proximity of the active volcanic zone.

An attempt is made to visualize the changes that have occurred within the barren communal grazing areas in Fig. 12.29. It shows parts of two communal areas in NE Iceland, divided by the glacial river Skjálfandafljót at the

Fig. 12.29 Now and then. A satellite image showing the current vegetation cover (on the *left*) and possible vegetation cover about 200 years after the Settlement (on the *right*) in two communal areas in NE Iceland. The river is the glacial river Skjálfandafljót. The vegetation cover on the right has been added to the original image. There is a dramatic difference between the current state and the expected vegetation cover in the past. The area can be restored to the former condition, but a long time is required (possibly 50–200 years, depending on inputs such as fertilizers and planting) with initial protection from grazing. Image processing based on SPOT (© EuroImage) by AUI/SM. © OA/SM; AUI

center with a perspective looking inland (south). The figure on the left shows the current condition with scattered vegetation patches in depressions. A road winds into the interior on the right (west side of the river). The two commons shown on the figure are among the poorest vegetated commons, but both are still used for grazing. On the figure on the right, an attempt has been made to add vegetation cover to the land, to resemble the condition about 200 years after the Settlement. The area is almost fully vegetated with some erosion on the steepest parts of the landscape. Birch forest dominates the valleys. This land could support a high number of grazing animals without damaging the environment, with good grazing practices. The difference is quite dramatic.

The Soil Conservation Service has protected many severely degraded areas from grazing, but there is a growing pressure to return grazing on some of these areas, even though they are in a poor state and mostly deserts. This shows the need for improving environmental education and communication in Iceland, and the need to tailor environmental law more to ensure protection and the restoration of severely degraded ecosystems.

12.6 Erosion Control and Restoration Perspectives

12.6.1 Early Efforts and Drivers for Restoration

Realization of the severe degradation and soil erosion was part of the "Enlightment" of the late nineteenth century, with advancing sand fronts being a constant threat to many communities. Organized efforts began in the Westfjords in the late eighteenth century by the priest Björn Gunnlaugsson, where calcareous beach sand was advancing into a small agricultural valley. There are accounts of sand burying farms in NE Iceland from the late eighteenth century (Crofts 2011). Earlier attempts to halt sand drift are manifested in old structures to halt sand drift, but they are poorly accounted for (Crofts 2011). Encroachment of sand from the Thjorsá river and the Hekla tephra deposits caused a massive destruction of one farm after the other in the late nineteenth and the early twentieth centuries in South Iceland. In 1907, while Iceland was under Danish rule, an organization was established to halt the sand and to protect and re-establish the tree cover. The Danish had successfully halted sand drift and re-established tree cover in severely degraded areas in Denmark (see Sigurjónsson 1958; Arnalds 1988; Olgeirsson 2007; Crofts 2011). There were only small patches of tall birch forests remaining in Iceland at that time, which were saved at the "last minute." At that time, Iceland was among the poorest nations anywhere on Earth, with rapid population growth, scarcity of food and fuel, and escalating land degradation problems. This organization that was established in 1907 developed into the current Soil Conservation Service (initially as the Sand Reclamation Service) and the Forest Service of Iceland. Thus, Iceland hosts one of the world's oldest operating national Soil Conservation agencies, while most other national agencies where established after the US Soil Conservation Service was initiated in the 1930s.

Halting the sand drift was the initial driver of the restoration work in Iceland as reviewed in a recent paper by Aradottir et al. (2013). The early efforts to halt the sand encroachment gradually became more and more successful during the twentieth century. Lyme grass (*Leymus arenarius*), a native sand-tolerant species, was (and still is) widely used to stabilize the sand, but various materials such as rocks and corrugated iron were used to raise wind brakes (Olgeirsson 2007).

Halting soil erosion was a major emphasis of the Soil Conservation up to the end of World War II, but after the war there was a growing emphasis on establishing new vegetation areas to enhance agricultural production, mainly grazing areas for sheep (Magnusson 1997; Petursdottir and Aradottir 2011). The latter part of the twentieth century saw increasing emphasis on "paying the debt to the land" (idealistic reasons for restoration work), but ecological restoration and nature conservation are currently growing as drivers for reclamation projects in Iceland (Aradottir et al. 2013). Other drivers include carbon sequestration in soils and vegetation to meet national commitments in relation to the UN Framework Convention on Climate Change (Petursdottir and Aradottir 2011; Aradottir et al. 2013).

12.6.2 Revegetation—Restoration

The terms "restoration" and "ecological restoration" have been defined in several different ways (see Aradottir and Hagen 2013 about restoration concepts and definitions) but the terms can be perceived as actions to facilitate the recovery of damaged or degraded ecosystems, to similar ecological function as existed before disturbance or degradation of a system. The terms "reclamation" and "revegetation" have more general meanings and involve actions to establish vegetation cover and ecological function on the surface. All these terms have frequently been used in relation to efforts to halt erosion and/or to restore degraded land and deserts in Iceland to productive ecosystems. Icelandic has a unique term "landgræðsla" which literarily means "landhealing," which is used in the Icelandic name of the Soil Conservation Service ("Landgræðslan"). This term is often used for all actions taken to improve the condition of the land, including protecting degraded lands from grazing. The Icelandic term for ecological restoration is "vistheimt."

Aradottir and Halldorsson (2011) edited a comprehensive volume on ecological restoration in Iceland ("Vistheimt á Íslandi," in Icelandic), including a review of restoration projects and the history of restoration efforts. The history of soil conservation in Iceland, including halting soil erosion and reseeding was published by Olgeirsson (2007) in Icelandic, rewritten in part in English by Crofts (2011). Other overviews available in English include those of Runolfsson (1987), Arnalds et al. (1987), Magnusson (1997), Aradottir and Eysteinsson (2005), and Aradottir et al. (2013).

There is considerable research that has been carried out over the years in relation to restoration efforts in Iceland. A long list of research projects with about 250 cited references was published by Aradottir et al. (2011) in the book "Vistheimt á Íslandi" (Ecological Restoration in Iceland). Early research focused on fertilizer experiments and finding suitable grass species for seeding in desert areas, both native and introduced, first with the intention to create fodder for sheep, but later increasingly to provide initial surface stability to allow natural establishment and succession of native species (see Magnusson 1997; Aradottir et al. 2013). Succession was studied on the newly formed Surtsey volcanic island (e.g., Kristinsson and Heidmarsson 2009; Magnusson et al. 2009), in lavas (e.g., Cutler et al. 2008), and on recently formed moraines in front of retreating glaciers (e.g., Marteinsdottir et al. 2010). A growing body of research in relation to reclamation and restoration has emerged over the past 30 years (Arnalds et al. 1987; Elmarsdottir et al. 2003; Gretarsdottir et al. 2004). These studies are not limited to vegetation cover, but also deal with soils and factors such as arthropods (e.g., Oddsdottir et al. 2008). New ecological methods to restoration involving sward and turf transplants have recently been studied at the Agricultural University of Iceland (e.g., Aradottir 2012; Aradottir and Oskarsdottir 2013).

Multidisciplinary research in relation to reclamation projects was initiated in the 1990s on the Blanda reclamation areas in NW highlands, where barren areas were being reclaimed to substitute foodstuff for grazing when a large hydroelectric reservoir was created in the area (see Arnalds et al. 1987; Thorsteinsson 1991; Gudmundsson 1991). More recent large reclamation and restoration ecology research projects, with specific studies of soil parameters include the LandAid and CarbBirch projects and studies by the Icelandic Soil Conservation Service on carbon sequestration within reclamation areas. The LandAid research area is unique for its extent, with 40 research plots, each being 1 ha, located within a 350 ha area dedicated to this research (Fig. 12.30). The research was initiated in 1999 with 10 commonly applied reclamation treatments, replicated four times (see Arnalds et al. 2013).

12.6.3 Reclamation and Soil Development

An important element of restoring ecosystems is renewing soil functions, such as organic carbon content, nitrogen and other nutrient levels, water infiltration, water holding capacity, and reducing evaporation. The results from these reclamation studies show in general a reversing of the trends

Fig. 12.30 The Geitasandur Landaid Restoration Research Area, infrared satellite image showing vegetation as *red*. The 1 ha plots show up as *squares* on the image. *Satellite image* © CNES/SPOT Image Corporation 2003

for the soil and surface factors in Table 12.3. Carbon accumulates in the soil at a relatively rapid rate of 0.04–0.08 kg C m^{-2} $year^{-1}$, owing to the andic nature of the soils, fast weathering rates of the basaltic tephra, and in part cold climate (Aradottir et al. 2000; Arnalds et al. 2002, 2013). This rate can be maintained for >100 years. Research shows significant increase in carbon with repeated measurements few years apart (Arnalds et al. 2013). One of the most notable change with restoration of the desert systems is a drop in pH from >7 to 5.5–6.5 in dryland systems. Clay formation takes longer time than the carbon accumulation (>50 to 100s of years to make significant changes). The organic matter increase results in improved water storage. However, the most dramatic changes occur on the surface, where vegetation that includes biological soil crust hinders surface movements by erosion and needle-ice (e.g., Oskarsson and Brynleifsdottir 2009). Infiltration in winter also improves gradually (Fig. 12.31; Orradottir et al. 2008; see also Strachan et al. 1998 for older restored soils). Temperature extremes and evaporation are reduced (Fig. 12.32).

Fig. 12.31 Measuring infiltration rates at the Geitasandur LandAid research cite. This an untreated control plot, with rapid infiltration in summer but very limited infiltration in winter time when the soil is frozen

Nitrogen accumulation is one of the most limiting factors after initial stabilization of the surface and the establishment of native species. As the deserts are almost devoid of nitrogen, the pool has to be accumulated from "scratch." Nitrogen deposition from the atmosphere in Iceland is only about 1 kg ha^{-1} $year^{-1}$. Studies that involve carbon accumulation for >50 years following restoration show that 20–40 kg N are accumulated each year in some lowland areas (which include 2 × 100–150 kg fertilizer applications), but much slower at higher elevations, from biological nitrogen fixation of soil crusts and other organisms in areas that are not grazed. Fully vegetated systems have, however, >4,500 kg N in the top 30 cm on average (AUI unpublished data). International studies (see Whisenant 1999) show that a minimum of 750–1,000 kg N is needed to reach adequate levels for ecosystem functions. Thus, it takes decades or even >100 years at higher elevations to reach sufficient nitrogen levels for robust ecosystem functions.

Grazing slows the accumulation rates in the early successional stages substantially and can halt it altogether. Grazing has also very detrimental effect on biological soil crusts which are key "players" in stabilizing the surface and

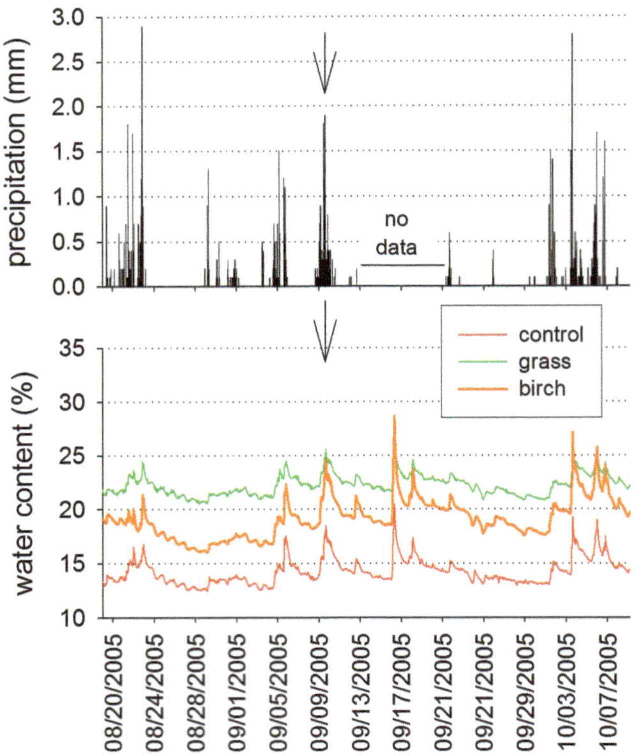

Fig. 12.32 Precipitation events and water contents at 5 cm depth in control (desert), and 5-year revegetated plots with grass/fertilizer and grass/fertilizer/birch treatments. The water content remains consistently lower in the control plots

accumulating the nitrogen. It is therefore important to prevent grazing in desert areas in Iceland in addition to multiple other factors that call for such intervention.

12.6.4 Many Ecosystems Are Being Reclaimed

Extensive areas have been revegetated in Iceland, and in many parts of the country, the vegetation cover is improving due to less or no grazing pressure on previously heavily grazed land. Much of the sandy deserts of the lowlands south and southwest of Mt. Hekla now have vegetation cover where barren sand was before. Many glacial rivers in South and Southeast Iceland have been channelized, reducing the extent of summer flooding and allowing for restoration and agricultural practices. Reclamation activities include seeding of grasses, applying fertilizers, and planting areas with trees (mostly by the Icelandic Forest Service but also by NGO's, farmers and other private landowners). Protecting from grazing is an important part of the land-healing activities, with 1,500 km^2 protected from grazing and additional 2,000 km^2 with partial protection and active land treatment in places by the Icelandic Soil Conservation Service (Halldorsson et al. 2011). The Icelandic Soil Conservation estimates that 571,000 ha (5,710 km^2) of land has been reclaimed in relation to the activities of the institute since 1907 (Crofts 2011). Halldorsson et al. (2011) listed all major restoration projects of the ISCS (in Icelandic).

Grazing practices have changed over the past decades. Horses are no longer grazed on the highland commons (with few exceptions), and the communal grazing period of sheep is becoming shorter. The number of sheep peaked around 1980 was nearly 1,000,000 winterfed ewes (see Arnalds and Barkarson 2003), but is currently below 500,000, which yet is much higher than the number during all the Middle Ages. Many sheep farms have been abandoned during the past decades, which have resulted in areas that now have limited grazing pressures in contrast to earlier days. Many of these areas are becoming greener with barren patches slowly becoming vegetated (e.g., around Reykjavík, in West Iceland, parts of the South, Southwest, West, and in the southwest part of the Westfjords). Reclamation and forestry activities are also slowly changing the Icelandic landscape toward more green, but almost exclusively in lowland areas. In some areas, more biological production is enhanced by a warm period, which can be related to global climate change (Hanna et al. 2004). The changes in green biomass show up in comparisons of satellite images 1982–2010 (Icelandic Institute of Natural History: http://www.ni.is/frettir/nr/13534).

The Icelandic Soil Conservation Service is involved in numerous land reclamation projects in co-operation with farmers and NGO's. The largest single project is the Hekluskógar project (Hekla Woodlands Project), which is intended to re-establish forests on the deserts around Mt. Hekla (Aradottir 2007; Oskarsson 2011). The project area is about 900 km^2 and a proportion of the area has very unstable ash and pumice surface. Much of the area was previously forested (birch and willows), but volcanic eruptions, wood harvesting, and continuous grazing (prevents regeneration of trees) have caused catastrophic erosion and desertification in the area. Woodlands are capable of receiving and stabilizing large amount of tephra, and the goal is to create resilient and stable ecosystems that can challenge inevitable future eruptions in the volcano. The method involves the use fertilizers and grasses to stabilize the surface and the establishment of islands of birch trees. The grasses give in, but provide safe sites for the birch to colonize from seed dispersal and subsequently spread further from clusters planted specifically for seed generation (Oskarsson 2011).

"Farmers Heal the Land" is a participatory approach project run by the Icelandic Soil Conservation Service, which involves about 15,000 ha (150 km^2) of land (see Arnalds 2005, 2011; Petursdottir 2011). The ISCS provides the fertilizer and guidance about land use and land reclamation. The areas involved are often used for grazing over

Fig. 12.33 The Þórsmörk area, a 4,000 ha restoration area, where birch, willows, and other natural vegetation have gained stronghold after protection from grazing, but the most severe erosion areas were fertilized and seeded for stabilization. The area in the foreground was barren soil until recently. Þórsmörk is now very popular for recreation activities. *Photo* © Gudjon Magnusson (Soil Conservation Service)

short periods, e.g., during the fall after full summer vegetation cycle. Large proportion of Icelandic farmers (about 650) are enrolled in the program (Petursdottir 2011).

Protection from grazing is often a major action taken to initiate and follow through land restoration by natural succession. Þórsmörk (The Garden of Thor) in the neighborhood of Eyjafjallajökull and the volcano Katla, a beautiful area popular for recreation activities, is perhaps one of the most impressive restoration areas in Iceland. It is about 4000 ha area where soil erosion and wood harvesting had nearly destroyed all the soils and the forest around 1900 (Oskarsson et al. 2011). During the twentieth century, the area was gradually protected from grazing and the birch forest has spread again over much of the area (Fig. 12.33). The Þórsmörk area is an example what can be achieved in restoration of Icelandic ecosystems.

References

***Regarding punctuation and Icelandic characters in citations*: See note on punctuation in the Preface**

Aradottir AL (2007) Restoration of birch and willow woodland on eroded areas. In: Halldorsson G, Oddsdottir ES, Eggertsson O (eds) Effects of afforestation on ecosystems, landscape and rural development. TemaNord 508:67–74

Aradottir AL (2012) Turf transplants for restoration of alpine vegetation: does size matter? J Appl Ecol 49:439–446

Aradottir AL, Arnalds O (2001) Ecosystem degradation and restoration of birch woodlands in Iceland. In: Wielgolaski FE (ed) Nordic mountain birch ecosystems. Man and the biosphere series. UNESCO, Paris, and Parthenon Publishing, Carnforth, pp 295–308

Aradottir AL, Eysteinsson Th (2005) Restoration of birch woodlands in Iceland. In: Stanturf JA, Madsen P (eds) Restoration of boreal and temperate forests. CRC Press, Boca Raton, pp 195–209
Aradottir AL, Halldorsson G (eds) (2011) Ecological restoration in Iceland. Agricultural University of Iceland and Icelandic Soil Conservation Service, Reykjavik. ISBN:978-9979-881-08-7
Aradottir AL, Hagen D (2013) Ecological restoration: approaches and impacts on vegetation, soils and society. Adv Agron 120:173–222
Aradottir AL, Oskarsdottir G (2013) The use of native turf transplants for roadside revegetation in a subarctic area. Icel Agric Sci 26:59–67
Aradottir AL, Arnalds O, Archer S (1992) Degradation of vegetation and soils (Hnignun grodurs og jardvegs). Græðum Island, soil conservation service yearbook 1991–1992:73–82 (in Icelandic)
Aradottir AL, Svavarsdottir K, Jonsson ThH, Gudbergsson G (2000) Carbon accumulation in vegetation and soils by reclamation of degraded areas. Icel Agric Sci 13:99–113
Aradottir AL, Halldorsson G, Svavarsdottir K, Arnalds O (2011) Ecological restoration research; an overview (Vistheimtarrannsóknir – yfirlit). In: Aradottir AL, Halldorsson G (eds) (2011) Ecological restoration in Iceland. Agricultural University of Iceland and Icelandic Soil Conservation Service, Reykjavik. ISBN:978-9979-881-08-7, pp 112–127 (in Icelandic)
Aradottir AL, Petursdottir T, Halldorsson G, Svavarsdottir K, Arnalds O (2013) Drivers of ecological restoration: lessons from a century of restoration in Iceland. Ecology and Society 18:33. http://dx.doi.org/10.5751/ES-05946-180433
Archer S, Stokes CJ (2000) Stress, disturbance and change in rangeland ecosystems. In: Arnalds O, Archer S (eds) Rangeland desertification. Kluwer Publishing Company, Dordrecht, pp 17–38
Arnalds A (1987) Ecosystem disturbance in Iceland. Arct Alp Res 19:508–513
Arnalds A (1988) Landgæði á Íslandi fyrr og nú (land resources in Iceland, past and present). Græðum Ísland (Icelandic soil conservation yearbook) 1:13–32
Arnalds A (2005) Approaches to landcare—a century of soil conservation in Iceland. Land Degrad Dev 16:113–125
Arnalds A (2011) Farmers heal the land: a social license for agriculture in Iceland. In: Williams J, Martin P (eds) Defending the social license of farming. CSIRO Publishing, Canberra, pp 83–92
Arnalds O (1990) Characterization and erosion of andisols in Iceland. Unpublished PhD dissertation, Texas A&M University, College Station, Texas
Arnalds O (1997) To read the land. Agricultural Research Institute (now AUI) and the Soil Conservation Service, Reykjavik. ISBN:9979-60-307-0 (in Icelandic)
Arnalds O (2000) The Icelandic "rofabard" soil erosion features. Earth Surf Proc Land 25:17–28
Arnalds O (2010) Dust sources and deposition of aeolian materials in Iceland. Icel Agric Sci 23:3–21
Arnalds O (2013) The influence of volcanic tephra (ash) on ecosystems. Adv Agron 121:331–380
Arnalds O, Barkarson BH (2003) Soil erosion and land use policy in Iceland in relation to sheep grazing and government subsidies. Environ Sci Policy 6:105–113
Arnalds O, Aradottir AL (2011) Rask á Íslandi (Ecosystem disturbance in Iceland). In: Aradottir AL, Halldorsson G (eds) Vistheimt á Íslandi (restoration in Iceland). Agricultural University of Iceland and the Icelandic Soil Conservation Service, Reykjavik, pp 19–23
Arnalds O, Aradottir AL (2014) To read and heal the land. Icelandic Environmental Association (Landvernd) and the Icelandic Soil Conservation Service, Reykjavik (in Icelandic)
Arnalds O, Aradottir AL, Thorsteinsson I (1987) The nature and restoration of denuded areas in Iceland. Arct Alp Res 19:518–525
Arnalds O, Thorarinsdottir EF, Metusalemsson S, Jonsson A, Gretarsson E, Arnason A (1997) Jardvegsrof á Íslandi (Soil Erosion in Iceland). The Icelandic Soil Conservation Service and the Agricultural Research University (now Agricultural University of Iceland), Reykjavik (in Icelandic). http://jardvegsstofa.lbhi.is/lisalib/getfile.aspx?itemid=1806
Arnalds O, Thorarinsdottir EF, Metusalemsson S, Jonsson A, Gretarsson E, Arnason A (2001) Soil erosion in Iceland. The Soil Conservation Service (Landgræðsla ríkisins) and the Agricultural Research Institute (Rala), Reykjavík. Translation of the original 1997 publication. http://jardvegsstofa.lbhi.is/lisalib/getfile.aspx?itemid=1807
Arnalds O, Aradottir AL, Gudbergsson G (2002) Organic carbon sequestration by restoration of severely degraded areas in Iceland. In: Kimble JM, Lal R, Follet RF (eds) Agricultural practices and policies for carbon sequestration in Soil. Lewis publishers, CRC Press, Boca Raton, pp 267–280
Arnalds O, Orradottir B, Aradottir AL (2013) Carbon accumulation in Icelandic desert Andosols during early stages of restoration. Geoderma 193–194:172–179
Arnalds O, Olafsson H, Dagsson-Waldhauserova P (2014) Quantification of iron-rich volcanogenic dust emissions and deposition over ocean from Icelandic dust sources. Biogeosci Discuss 11:5941–5967
Bestelmeyer BT, Goolsby DP, Archer SR (2011) Spatial perspectives in state-and-transition models: a missing link to land management? J Appl Ecol 48:746–757
Bjarnason H (1942) Ábúð og örtröð (Land use and over-exploitation). Icel Forest Soc Yearb 1942:8–40 (in Icelandic)
Casanova M, Salazar O, Seguel O, W Luzio W (2013) The Soils of Chile. World soils book series. Springer, Dordrecht
Crofts R (2011) Healing the land. Soil Conservation Service of Iceland, Gunnarsholt. ISBN 978-9979-9295-5-0
Cutler NA, Belyea LR, Dugmore AJ (2008) The spatiotemporal dynamics of primary succession. J Ecol 96:231–246
Dagsson-Waldhauserova P, Arnalds O, Olafsson H (2013) Long-term frequency and characteristics of dust storm events in Northeast Iceland (1949–2011). Atmos Environ 77:117–127
Diamond J (2005) Collapse. Penguin, New York
Doner L (2003) Late-Holocene paleoenvironments of northwest Iceland from lake sediments. Palaeogeogr Palaeoclimatol Palaeoecol 193:535–560
Dugmore AJ, Newton AJ, Larsen G, Cook GT (2000) Tephrochronology, environmental change and the Norse Settlement of Iceland. Environ Archaeol 5:21–34
Dugmore AJ, Church MJ, Mairs K-A, McGovern TH, Perdikaris S, Vesteinsson O (2007) Abandoned farms, volcanic impacts, and woodland management: revisiting Þjórsárdalur, the „Pompeii of Iceland". Arctic Anthropol 44:1–11
Einarsson Th (1962) Vitnisburður frjógreiningar um gróður, veðurfar og landnám á Íslandi (Pollen evidence about vegetation, climate and the Settlement of Iceland). Saga 1962:442–469 (in Icelandic)
Einarsson Th (1968) Jarðfræði. Saga bergs og lands (Geology, a History of Rocks and Land). Mál og Menning, Reykjavík (in Icelandic)
Elmarsdottir A, Aradottir AL, Trlica MJ (2003) Microsite availability and establishment of native species on degraded and reclaimed sites. J Appl Ecol 40:815–823
Erlendsson E, Edwards KJ, Buckland PC (2009) Vegetational response to human colonisation of the coastal and volcanic environments of Ketilstaðir, southern Iceland. Quat Res 72:174–187
Gardarsson A, Magnusson B, Thorleifsson EO, Oskarsson H, Hilmarsson JO, Lund NA, Thrainsson S, Baldursson T (2006) Endurheimt votlendis 1996–2006. (Wetland restoration 1996–2006). Report of the Wetland Committee (Skýrsla Votlendisnefndar). Daniel Bergmann ed. Agricultural Ministry, Reykjavík (in Icelandic)
Geirsdottir A, Miller GH, Thordarson Th, Olafsdottir KB (2009) A 2000 year record of climate variations reconstructed from Haukadalsvatn, West Iceland. J Paleolimnol 49:95–115

Geirsson O (1998) Draining of wetlands. In: Olafsson JS (ed) Icelandic Wetlands. Háskólaútgáfan (University Publishing), University of Iceland, Reykjavik, pp 105–119 (in Icelandic)

Gerrard AJ (1985) Soil erosion and landscape stability in southern Iceland: a tephrochronological approach. In: Richards K, Arnett R, Ellis S (eds) Geomorphology and soils. George Allen and Unwin Ltd, UK, pp 78–95

Gisladottir FO, Arnalds O, Gisladottir G (2005) The effect of landscape and retreating glaciers on wind erosion in South Iceland. Land Degrad Dev 16:177–187

Gisladottir G (1998) Environmental characterisation and change in South-western Iceland. PhD thesis, The Department of Physical Geography, Stockholm University dissertation series no 10, Stockholm, Sweden

Gisladottir G, Erlendsson E, Lal R, Bigham J (2010) Erosional effects on terrestrial resources over the last millennium in Reykjanes, southwest Iceland. Quat Res 73:20–32

Gisladottir G, Erlendsson E, Lal R (2011) Soil evidence for historical human-induced land degradation in West Iceland. Appl Geochem 26:S28–S31

Gislason SR (2012) The Carbon Cycle (Kolefnishringrásin). Hið Íslenska Bókmenntafélag, Reykjavík. ISBN 978-9979-66-303-4

Greipsson S (2012) Catastrophic soil erosion in Iceland: impact of long-term climate change, compounded natural disturbances and human driven land-use changes. Catena 98:41–54

Gretarsdottir J, Aradottir AL, Vandvik V, Heegaard E, Birks JB (2004) Long-term effects of reclamation treatments on plant succession in Iceland. Restor Ecol 12:268–278

Gudbergsson G (1975) Soil Formation in Skagafjörður, northern Iceland. J Agric Res Icel 7:20–45 (in Icelandic, extended English summary)

Gudbergsson G (1996) The influence of human habitation on soil and vegetation in three counties in North-Iceland. Icel Agric Sci 10:31–81 (in Icelandic, extended English Summary)

Gudmundsson HJ (1997) A review of the Holocene environmental history of Iceland. Quat Sci Rev 16:81–92

Gudmundsson Th (1991) Soils of the experimental areas (Jarðvegsrannsóknir á tilraunasvæðum). In: Thorsteinsson I (ed) Reclamation on Auðkúluheiði and Eyvindarstaðaheiði 1981–1989. Rala report 151, Agricultural University of Iceland, Reykjavík, pp 51–70 (in Icelandic)

Halldorsson G, Svavarsdottir K, Thorarinsdottir EF, Runolfsson S (2011) Ecological restoration by the Soil Conservation Service, pp 40–48 (in Icelandic)

Hallsdottir M (1995) On the pre-settlement history of Icelandic vegetation. Icel Agric Sci 9:17–29

Hallsdottir M, Caseldine CJ (2005) The Holocene vegetation history of Iceland, state-of-the art and future research. In: Caseldine C, Russel A, Hardardottir J, Kudsen O (eds) Iceland—modern processes and past environments. Dev Quat Sci 5:319–334

Hallsdottir BS, Wöll C, Gudmundsson J, Snorrason A, Thorsson J (2013) Emissions of greenhouse gases in Iceland from 1990 to 2011. National inventory report 2013. Environmental Agency of Iceland, Reykjavik

Hanna E, Jonsson T, Bos JE (2004) An analysis of Icelandic climate since the nineteenth century. Int J Climatol 24:1193–1210

Helgadottir A, Eythorsdottir E, Johannesson T (2013) Agriculture in Iceland—a grassland based production. Grassl Sci Eur 18:30–43

Herrick JE, Sala OE, Karl JW (2013) Land degradation and climate change: a sin of omission? Guest editorial. Front Ecol 11:213

Imeson A (2012) Desertification, land degradation and sustainability. Paradigms, processes, principles and policies. Wiley-Blackwell, Chichester

Jonsdottir IS (1984) Effects of grazing on the vegetation of Auðkúluheiði heathland, northern Iceland. Náttúrufræðingurinn 53:19–40 (in Icelandic)

Karlsson G (2000) Icelands's 1100 years. History of a Marginal Society. C. Hurst & Co Ltd, London

Krasilnikov P, del Carmen Gutiérrez-Castorena Ma, Ahrens RJ, Cruz-Gaistardo CO, Sedov S, Solleiro-Rebolledo E (2013) The soils of Mexico. In: World book series. Springer, Dordrecht

Kristinsson H, Heidmarsson S (2009) Colonization of lichens in Surtsey 1970–2006. Surtsey Res 12:81–104

Larsen DJ, Miller GH, Geirsdottir A, Thordarson Th (2011) A 3000-year varved record of glacier activity and climate change from the proglacial lake Hvítárvatn, Iceland. Quat Sci Rev 30:2715–2731

Lawson IT, Gathorne-Hardy FJ, Church MJ, Newton AJ, Edwards KJ, Dugmore AJ, Einarsson A (2007) Environmental impacts of the Norse settlement: palaeoenvironmental data from Mývatnssveit, northern Iceland. Boreas 36. doi:10.1080/03009480600827298

Magnusson B (1998) Vegetation of drained mires (Gróður í framræstum mýrum). In: Olafsson JS (ed) Icelandic Wetlands. Háskólaútgáfan (University Publishing), University of Iceland, Reykjavik, pp 105–119 (in Icelandic)

Magnusson B, Magusson SH, Fridriksson S (2009) Developments in plant colonization and succession on Surtsey during 1999–2008. Surstey Res 12:57–76

Magnusson SH (1997) Restoration of eroded areas in Iceland. In: Urbanska KM, Webb NR, Edwards PJ (eds) Restoration ecology and sustainable Dev. Cambridge University Press, Cambridge, pp 188–211

Marteinsdottir B, Svavarsdottir K, Thorhallsdottir Th (2010) Development of vegetation patterns in early primary succession. J Veg Sci 21:531–540

Montgomery DR (2007) Dirt. University of California Press, Berkeley

NRC (1993) Rangeland health. National Research Council. National Academic Press, Washington DC

Oddsdottir ES, Svavarsdottir K, Halldorsson G (2008) The influence of land reclamation and afforestation on soil arthropods in Iceland. Icel Agric Sci 21:3–15

Ogilvie AEJ (2005) Local knowledge and travellers' tale: a selection of climate observations in Iceland. In: Caseldine C, Russel A, Hardardottir J, Kudsen O (eds) Iceland—Modern processes and past environments. Dev Quat Sci 5:257–287

Ogilvie AEJ, Jonsson T (2001) "Little Ice Age" research: a perspective from Iceland. Clim Change 48:9–52

Olafsdottir S (2013) Reasons for and impact from wetland drainage in Mýrasýsla 1930–1990 (Ástæður og áhrif framræslu í Mýrasýslu). BS thesis, Agricultural University of Iceland, Hvanneyri, Borgarnes, Iceland

Olafsdottir R, Schlyter P, Haraldsson HV (2001) Simulating Icelandic vegetation cover during the Holocene. Implications for long-term land degradation. Geogr Ann 83A:203–215

Olgeirsson FG (2007) The Seeders of Sand. The history of land reclamation in Iceland 1907–2007. (Sáðmenn sandanna. Saga landgræðslu á Íslandi 1907–2007). Icelandic Soil Conservation Service, Gunnarsholt (in Icelandic)

Orradottir B, Archer SR, Arnalds O, Wilding LP, Thurow TL (2008) Infiltration in Icelandic Andisols: the role of vegetation and soil frost. Arct Antarct Alp Res 40:412–421

Oskarsson BV, Riishuus MS, Arnalds O (2012) Climate-dependent chemical weathering of volcanic soils in Iceland. Geoderma 189–190:635–651

Oskarsson H (1998) Wetland drainage of the western lowlands (Framræsla votlendis á Vesturlandi). In: Olafsson JS (ed) Icelandic wetlands. Háskólaútgáfan (University Publishing), University of Iceland, Reykjavik, pp 121–129

Oskarsson H (2008) Wetland restoration—a potential method to reduce green-house gas emissions. Agricult Congr (Fræðaþing landbúnaðarins) 2008:298–301 (in Icelandic)

Oskarsson H (2011) The Hekla woodlands project (Hekluskógar). In: Aradottir AL, Halldorsson G (eds) Vistheimt á Íslandi (restoration in Iceland). Agricultural University of Iceland and the Icelandic Soil Conservation Service, Reykjavik, pp 71–74 (in Icelandic). ISBN:978-9979-881-08-7

Oskarsson H, Brynleifsdottir SJ (2009) The interaction of fertilization in nursery and field on survival, growth and the frost heaving of birch and spruce. Icel Agric Sci 22:59–68

Oskarsson H, Magnusson G, Halldorsson G (2011) Thorsmörk and Godaland. In: Aradottir AL, Halldorsson G (eds) Ecological restoration in Iceland. Agricultural University of Iceland and Icelandic Soil Conservation Service, Reykjavik, pp 75–76 (in Icelandic). ISBN:978-9979-881-08-7

Petursdottir Th (2011) Farmers Heal the Land. In: Aradottir AL, Halldorsson G (eds) Ecological restoration in Iceland. Agricultural University of Iceland and Icelandic Soil Conservation Service, Reykjavik, pp 84–85 (in Icelandic). ISBN:978-9979-881-08-7

Petursdottir Th, Aradottir AL (2011) History of ecological restoration in Iceland. In: Aradottir AL, Halldorsson G (eds) Ecological restoration in Iceland. Agricultural University of Iceland and Icelandic Soil Conservation Service, Reykjavik, pp 31–36 (in Icelandic). ISBN:978-9979-881-08-7

Pierson FB (2000) Erosion models: use and misuse on rangelands. In: Arnalds O, Archer S (eds) Rangeland Desertification. Kluwer Academic Publishers, Dordrecht, pp 67–76

Runolfsson S (1987) Land reclamation in Iceland. Arct Alp Res 19:514–519

Sigbjarnarson G (1969) The loessial soil formation and the soil erosion on Haudadalsheidi. Náttúrufræðingurinn 39:68–118 (in Icelandic, extended English summary)

Sigurdsson S (1977) Birki á Íslandi (útbreiðsla og ástand). (Birch in Iceland—distribution and condition). In: Gudmundsson H (ed) Skógarmál. Reykjavik, Edda, pp 146–172 (in Icelandic)

Sigurjonsson A (1958) Ágrip af gróðursögu landsins til 1880 (A summary of the vegetation history of Iceland to 1880). In: Sigurjonsson A (ed) Sandgræðslan. Minnzt 50 ára starfs Sandgræðslu ríkisins. Búnaðarfélag Íslands and Sandgræðsla ríkisins, Reykjavík (in Icelandic)

SRM (1992) Society of range management position statements. Trailboss News, Dec 1992:5

Stoops G, Gerard M, Arnalds O (2008) A micromorphological study of Andosol genesis in Iceland. In: Kapur S, Mermut A, Stoops G (eds) New trends in micromorphology. Springer, Heidelberg, pp 67–90

Strachan IB, Sigurdsson BD, McCaughey JH (1998) Soil hydrology at the Gunnarsholt plantation: measurements and results. Icel Agric Sci 12:27–34

Streeter RT, Dugmore AJ (2012) Reconstructing late-Holocene environmental change in Iceland using high-resolution tephrochronology. Holocene 23:197–207

Streeter R, Dugmore AJ, Vesteinsson O (2012) Plague and landscape resilience in premodern Iceland. PNAS 109:3664–3669

Sveinbjarnardottir G, Buckland PC, Gerrard AJ (1982) Landscape change in Eyjafjallasveit, southern Iceland. Norsk Geografisk Tidskrifter 36:75–88

Thorhallsdottir ThE, Thorsson J, Sigurdardottir S, Svavarsdottir K, Johannsson MH (1998) Wetland disturbance in the southern lowlands (Röskun votlendis á Suðurlandi). In: Olafsson JS (ed) Icelandic wetlands. Háskólaútgáfan (University Publishing), University of Iceland, Reykjavik, pp 131–142 (in Icelandic)

Thorarinsson S (1944) Tefrokronologiska studier på Island. Munksgaard, Copenhagen

Thorarinsson S (1979) On the damage caused by volcanic eruptions with special reference to tephra and gases. In: Sheets PD, Grayson DK (eds) Volcanic activity and human ecology. Academic Press, New York, pp 125–159

Thorsson J (2008) Desertification of high latitude ecosystems: Conceptual models, time-series analyses and experiments. Unpublished PhD dissertation, rangeland ecology and management, Texas A&M University, College Station, Texas

Thorsteinsson I (ed) (1991) Reclamation on Auðkúluheiði and Eyvindarstaðaheiði 1981–1989. Rala report 151, Agricultural University of Iceland, Reykjavik (in Icelandic)

Thorsteinsson I, Olafsson G, Van Dyne GM (1971) Range resources of Iceland. J Range Manag 24:86–93

Valdimarsdottir AM, Magnusson SH (2013) Vegetation of Videy, an isolated island in the River Thorsa. Náttúrufræðingurinn 83:49–60 (In Icelandic, English summary)

Vickers K, Erlendsson E, Church MJ, Edwards KJ, Bending J (2011) 1000 years of environmental change and human impact at Stóra-Mörk, southern Iceland. A multiproxy study of a dynamic and vulnerable landscape. Holocene 21:979–995

Wald EC (2012) Land-use development in South Iceland 1900–2010. MSc thesis, Faculty of Biology and Environmental Science, University of Iceland, Reykjavík

Whisenant SG (1999) Repairing damaged wildlands. A process orientated, landscape-scale approach. Biological conservation, restoration, and sustainability, vol 1. Cambridge University Press, UK

Wilson T, Cole JW, Stewart C, Cronin SJ, Johnston DM (2011) Ash storms: impacts of wind-remobilised volcanic ash on rural communities and agriculture following the 1991 Hudson eruption, southern Patagonia, Chile. Bull Volcanol 73:223–239

Index

O. Arnalds, *The Soils of Iceland*,
World Soils Book Series, DOI 10.1007/978-94-017-9621-7
© Springer Science+Business Media Dordrecht 2015

FSC
www.fsc.org
MIX
Papier | Fördert gute Waldnutzung
FSC® C083411

Kolibri 360 GmbH
Ferdinand-Jühlke-Straße 7
99095 Erfurt, Deutschland
produktsicherheit@kolibri360.de